ELECTRIC SAVVY
Using Electricity to Make Your House a Home

ELECTRIC SAVVY

Using Electricity to Make Your House a Home

Blaine C. Readler

ELECTRIC SAVVY
Using Electricity to Make Your House a Home

Copyright © 2016 by Blaine C. Readler

Visit us at: http://www.readler.com

E-mail: blaine@readler.com

ISBN: 978-0-9834973-6-3

Printed in the United States of America

This book is dedicated to Sam Parri, who taught me that without the prerequisite of work, excellence is just an idea.

ACKNOWLEDGEMENTS

Jennifer Silva Redmond has been editing my books for some time, and to say that she's improved them is to say that adding wings to the Wright brothers first airplane improved its ability to fly.

www.jennyredbug.com

Thanks to Jackson Finley for visualizing the essence of the book, and rendering it in such appealing pixels.

www.jwfinley.com

To cross the seas, to traverse the roads. and to work machinery by galvanism, or rather electro-magnetism, will certainly, if executed, be the most noble achievement ever performed by man.
Alfred Smee, 1841

I've found out so much about electricity that I've reached the point where I understand nothing and can explain nothing.
Pieter van Musschenbroek, ~1730,
describing his experiments with his invention, the Leyden jar

If it weren't for electricity, we'd all be watching television by candlelight.
George Gobel, 1954

Contents

INTRODUCTION

This book was written for those who never had much interest in how electricity worked in their everyday lives, until now. Perhaps you'd like to know more than just that electricity somehow keeps your refrigerator cold, but you're wary about drowning in the esoteric details of the physics involved. You don't want to *design* a refrigerator, you just want to understand the *basic* principles.

I've tried wherever possible to use concepts instead of quantities. This book includes only two equations, and leaving these out would be like describing football without mentioning yard lines. Instead of equations, I've attempted to convey a grasp of proportions, which, unless you're an engineer (and even then), is 95% of understanding the workings of electricity.

My goal is to remove the murky cloak of inscrutable jargon and unnecessary complexity and reveal the basic way that electricity animates your house and renders it a home. This is not a textbook. Although each chapter, in general, progresses from the previous, they don't necessarily build on each other, and it's not necessary to read them all. So if you find yourself yawning, go ahead and skip ahead. Or take a nap.

My favorite teachers in high school and college were the ones that welcomed humor. Besides the obvious effect of keeping things interesting, the casual approach lifted the curtain of apprehension on subjects that at first seemed daunting. In writing this book, I've followed their example. I'd like to say that I did it exclusively for you, but the truth is that it made the whole thing a lot more fun for me.

Speaking of daunting, there's a lot of footnotes. But, as you'll see, these are not dry academic references but, rather, interesting side notes—little tidbits and stories that would have interrupted the flow if inserted in vivo. Think of them as a friend on the movie crew making "Jurassic World" who slips you the 'inside poop' during refreshment breaks. If it weren't for footnotes, I wouldn't have discovered that Isaac Newton, considered by many the greatest physicist ever, had probed with a long sewing needle around the sides of his eyeballs, just to find out what was there.

I greatly enjoyed writing this book, I can only hope you have half as much fun reading it.

1
Beginnings

When my dad was a boy, he came home from school one day to find his mother (my grandmother), lying motionless on the floor of a small back room. She'd been doing laundry with the gas-powered washing machine, and had collapsed, a wet shirt still clutched in her hand. My dad got his older brother, who—suspecting what had happened— opened all the windows, letting in the freezing winter wind. My grandmother soon revived, promptly threw up, and was led off to bed, groggy, with a splitting headache. The gas engine exhaust hose had sprung a leak, and she'd succumbed to carbon monoxide poisoning. The doctor told her that if the boys had arrived ten minutes later, she'd be dead.

My ancestors were farmers and coal miners in rural Pennsylvania. Life was hard. The work days started, for man and woman, before the sun was up, and ended when the kerosene lanterns were extinguished and a last load of wood shoved into the stove where it would be cold ash by morning. Each February, my grandfather joined other farmers from around Wapwallopen to chop out blocks of thick ice from a frozen stream. They hauled the blocks to their ice houses— small buildings with double walls between which sawdust was packed for insulation. The ice blocks were piled inside, and then covered with layers of additional sawdust. With luck, the ice would last through the hot, muggy summers until the cool weather of autumn returned. Ice was precious, for it allowed meat to be stored and eaten over a span of days.[1]

Small gas engines become affordable by 1930, and were pressed into service wherever possible to save some labor. But there were inherent dangers, as my grandmother discovered. All the farmers had cars and tractors by this time—the days of horse and buggies had passed—but most had no electricity. Power utilities were slowly, grudgingly, expanding their distribution lines into the rural backwoods. My dad remembers walking along beneath the power lines, but the electricity wasn't free, and this was the Great Depression. One by one, the farmers scraped together the money to have their houses wired and to pay the utility connection fee. Herbie Sweet, a farmer and part-time electrician, did all the house wiring. Most of the old houses were plank construction—no inside space between studs—and so the wiring ran along baseboards, ceilings, and walls with a metal cover for protection. Not all rooms were wired—the kitchen came first, and then whatever else the farmer could afford.

As time passed, farmers extended their wiring on their own, overloading the couple of circuits Herbie had installed. Word got around that the problem of blown fuses could be circumvented by placing a penny behind the dead fuse. This meant, of course, that the original ten or fifteen amp fuse was now replaced with whatever a penny could conduct—hundreds of amps. Herbie cussed the farmers and told them the story about how his wife's hair got caught in the rollers of their new electric washing machine, and if the fuse hadn't blown, he'd now have a wife with no scalp. Whether the story was true or not, it convinced some farmers to remove their pennies.

---

[1] most farmers kept at least one milk cow, providing a ready supply daily. "Ready," requiring only daily milking, feeding, ministering, and shoveling the manure of the cow, which for its part just stood around all day digesting.

The first electric appliance to find its way into my grandparents' house was a radio, powered by large dry-cell batteries. After school, my dad listened to WJZ out of New York, broadcasting "Jack Armstrong—All American Boy," Ralston-Purina's "Tom Mix Straight Shooters," and, of course, "The Lone Ranger," and "Orphan Annie." News aired in the evening. Mornings, it was "Don McNeill's Breakfast Club" out of Chicago. This was their weather report, since whatever was happening over Illinois blew into northeast Pennsylvania a day or two later. Eventually the radio's big battery would die, and the house would go silent until another could be bought.

As with all the farms, when my grandparents' farm was eventually wired and connected, all they had were electric lights, bulbs screwed into ceiling sockets, and turned on and off by hanging cords. But, oh what a blessing. No more sooty, smelly kerosene lamps. No more fire hazards (as long as they resisted the penny solution).

As extra money became available, appliances followed, but sometimes not for years. For decades, even when I was a boy, the poles upon which were strung the electric cables (and later, telephone wires) were called "light poles."

The first device purchased by my grandparents that was made to plug into a wall socket was a radio. No more dead batteries, no more silent house. The world beyond Wapwallopen settled in as a permanent housemate. Next came a vacuum cleaner, followed by a washing machine.[2] My ancestry is mostly Pennsylvania Dutch (German), and cleanliness was elbow-to-elbow with godliness. My grandparents never had a refrigerator while my dad lived with them. This was a luxury for the well-to-do.

---

[2] this wasn't the kind of washing machine you're used to, where you toss in the clothes and detergent, spin the dial, and come back a half hour later to move them to the dryer. My mom pulled our juggernaut into the middle of the kitchen on washing day, and loaded the large tub with the week's clothes. A motor-driven agitator sloshed them around, but she used a stick to help it along, looking like a witch tending a brew. Once the large load was washed and rinsed, she swung around what looked like a gallows, but in place of the noose bar, were two belt-driven rubber rollers through which she would feed each article of clothing, squeezing out the excess water. Then outside to the clothes line to have them rained on in summer, and frozen in winter.

My dad doesn't remember any electrocutions during this time. Bathroom appliances—electric razors, curling irons, and blow dryers—were still in the future, and so the dangerous combination of high voltage and water was also waiting to shock future generations.

After my dad and mom married in 1950, the only electric appliance they owned for some months was a toaster. From the beginning, a toaster was the iconic American wedding present.[3]

∞

Prior to 1840, electricity was a mere scientific curiosity, sort of like what sub-atomic particles are today. News of battles won or lost took days or weeks to reach Washington, and then additional days or weeks for commands to be returned. The last battle of the War of 1812 was fought after the war was over; neither side knew that the Treat of Ghent had already been signed. The 440 casualties—closer to 800 if you include later deaths from infections, mainly from amputations—of the epic struggle to defend New Orleans (immortalized in song and legend), when Stonewall Jackson beat back the superior forces of the British, were lives lost for no reason other than a lack of technology.

In the last decade of the eighteenth century, the French and British developed a system of communication consisting of towers built atop hills a few miles apart that could display symbolic representations of alphabetical letters and numbers. Messages were received via telescope, then re-transmitted to the next tower down the line. At one point, over a thousand of these towers dotted the hills of France and England. This system, akin to the flag semaphore system still used by the Navy, was called a "telegraph," and worked well, compared to a man on a horse. When there was daylight. And no fog. Or rain.

---

[3] I was eight years old when my dad added a bathroom to the old farmhouse they'd bought. Until then, we used an outhouse. This was where I learned to read. Newspaper was used as toilet paper, and the comics went down the hole last. Not that any of this has anything to do with electricity, but I find it endearing.

But this "optical" telegraph was quite expensive for the governments, since each tower had to be manned continuously. By 1800 it was known by scientists that electricity traveled along a wire at near infinite speed, and many amateur inventors tried their hand at wrestling the phenomenon into service. Most attempts were misguided, but a few worked rudimentarily, if clumsily. The idea that communications could occur instantly and invisibly along a metal wire was far beyond the experience and ken of the rest of the population, however, and all these attempts were viewed as curious parlor tricks. Nobody took them serious, certainly not government officials who controlled funding that could produce working systems.

By the early 1830s, two amateur inventors—Samuel Morse in the United States, and William Cooke in England—independently developed methods of communicating individual letters and numbers using contact switches at the sending side of a wire and sensing electromagnets at the receiving side. In both cases, their nascent electrical telegraphs worked wonderfully, but only for a few dozen feet. Not a whole lot of utility in that. Each found a bona fide scientist to team with, though, and improved their electrical systems so that the communications could span miles. On both sides of the Atlantic, workable electrical telegraph methods were poised and waiting to change the world.

Unfortunately, about this time Alessandro Volta's "voltaic pile" batteries became generally available, and were eagerly taken up by the usual crowd of hucksters and swindlers. Electricity was sold as everything from a cure for any disease, to a means of communicating with the spirit world. The majority of the public were not fooled by these charlatans, but the consequence was that everything electric had become tainted. It was an uphill battle for both Morse and Cooke, but both eventually convinced their respective governments to fund experimental single-span systems along existing railroad tracks. Still, both the public and their representatives in Congress and Parliament considered the whole thing an interesting novelty. Religious leaders in Baltimore openly questioned whether the electrical telegraph wasn't a result of the black arts.

It took a few spectacular triumphs to break the ice—in England, the announcement in London of the birth of Queen Victoria's second son in Windsor an hour ahead of the first news arriving via train; in Baltimore, the identities of the nominees selected by the Whig national convention in Washington; and in both countries, the arrests of notorious criminals made possible only through the aid of instantaneous communication and the criminals' ignorance that electricity travels faster than a getaway train. Events such as these received much heralded coverage in the newspapers. Then, as now, the public media decides what, out of all that's new, we embrace.

Even then, the world wasn't quite ready to open their wallets and rush to the telegraph station with a message. Both Morse and Cooke worked on diligently for years, arranging business partners and finagling funding before the spreading network of wires took hold in the public's imagination and became an integral part of western civilization. Once the transcontinental telegraph line reached Salt Lake City, it spelled the end of the Pony Express, after just a year and a half of that romantic equestrian operation.

The success of the telegraph opened inventor's eyes to the practical promise of electricity. Pioneers like Alexander Graham Bell, Elisha Gray, Thomas Edison, and Joseph Swan found a medium to express their creative genius.[4] By the end of the nineteenth century, innovators of electricity were primed to change the lives of everyday people with electricity, just as industry had blossomed at the beginning of the

---

[4] you've probably never heard of Elisha Gray or Joseph Swan. Elisha was in a race with Bell to patent the first workable telephone, and by a bit of luck, Bell won. Swan was the inventor of the incandescent light bulb, Thomas Edison merely found an improved filament, and this was replaced by a better one—tungsten—twenty-five years later. Edison did for the light bulb, what George Eastman (of Kodak fame) did for the camera. I'll bet you were taught that Edison invented the light bulb—pure American hubris. History remembers and celebrates both Bell and Edison partly because they were ingenious inventors, but equally because they were willing to work against daunting odds to promote and develop their creations, just like Morse and Cooke.

century with the development of the steam engine. The telephone was about to connect people in ways the human race had never imagined. But both the telephones and the telegraph system before it still relied on the same voltaic pile batteries wielded with deleterious intent by the con men. To fully unfold the promise of electricity, the world needed a reliable, steady supply of electrical power, and for this, the world needed a go-getter bulldog to get it all started. This sort of thing is where the true genius of Thomas Edison lay, as we'll see in the next chapter.

As purported,[5] men of long vision at the 1893 World's Fair held in Chicago, greatly impressed by Alexander Graham Bell's demonstration of the telephone, predicted that someday there would be one in every city. One wonders how their waxed Victorian mustaches would have twitched had they known that one day there would be one embedded in every ear.

---

[5] by passages repeated verbatim on multiple websites. I can't remember the actual credible source where I first read this.

2

Feeding the House

Every modern house has two basic sources of electricity: the 120 volt AC provided by the utility company, and a variety of device batteries (AAA, AA, 9V, button, rechargeable, etc.). The main difference between the two is that the first is about four thousand times cheaper per watt than the second. But, economics aside, the utility company delivers AC, while the batteries are DC. You probably already know that AC stands for alternating current, and DC stands for direct current.

The electricity provided to the first homes over a century ago was actually DC. After Edison developed the first economical light bulb, he needed a market, and to that end, he developed the first electricity-generating utilities to service homeowners (so they could buy his light bulbs). Despite his genius in so many other areas, he blundered here by basing his utilities on direct current supply. DC electrical power is fine for short distances, but suffers losses over even moderate distances—measured in blocks, not miles. As we will see later, AC electricity has the advantage that its voltage level can be transformed up, and the higher the voltage, the more efficient its transport. This is obvious in the high-voltage lines—nearly a million volts in some cases—we find crisscrossing the nation. For home distribution, the high voltage is transformed back down to 120 volts. You may have seen these distribution stations, huddles of giant metal cans with ribbed layering (for cooling) surrounded by barbed-wire fencing (for your safety). Because of its transport efficiency, utilities generating AC eventually won out over Edison's DC versions. But if it wasn't for his bulldog determination to get the ball rolling, it would have been many years before the substantial initial investment hurdle was surmounted.

We still carry a legacy of Edison's early DC power utilities. A voltage of 120 volts may seem an odd, or perhaps arbitrary, value. In developing his commercial light bulb, he settled on 100 volts as a working value. This was a compromise between the relatively high voltage required for his carbon filaments, but not so high as to be (or so it was thought) dangerous. But because of DC's transport losses, he had to up the voltage to 120 volts, so that—after losses—a full 100 volts could be delivered to the customer. The AC-based utilities that eventually took over his territories had to accommodate the existing customers' bulbs, and so they followed suit. However, since AC experiences so little loss, the original 120 volts ends up in your AC outlets.

Now is as good a time as any to compare voltage and current. Fundamentally they are related in the same way that the gas pedal of your car and its speed are related— increase the throttle (the voltage) and the car goes faster (more current). A common analogy, and a good one, is water through a hose. Voltage is like the water pressure provided by the city, and the flow of water is like the current. Double the water pressure, and you get double the amount of water out of the hose. The actual amount of water the hose delivers is dependent on the size of the hose. This is like the resistance of an electrical circuit, where a smaller hose corresponds to more resistance.

When we increase voltage, the current increases proportionally. If we increase the resistance, the current decreases proportionally. The amount of voltage and resistance determines the amount of current. This is a very simple and direct relationship. So simple, in fact, that we'll introduce our first of only two equations:

I=V/R

The "I" stands for current measured in amps,[1] "V" for voltage measured in volts, and "R" for resistance measured in ohms. If we apply one volt across one ohm, we'll get one amp. If we apply one volt across ten ohms, we'll get one-tenth of an amp.

We can rearrange this equation like so:

V=I*R

If I know that the current is one amp, and that the resistance is one hundred ohms, then I know that the voltage being applied is one hundred volts.

Finally, I can solve for resistance:

R=V/I

If I know that ten volts is being applied, and the current is five amps, then I know that the resistance is two ohms.

These relationships (really just one relationship rearranged) are called Ohm's law, named after the German physicist Georg Simon Ohm, who noticed the relational association between voltage, current, and resistance and published it as part of an overarching book in 1827. His discovery wasn't much appreciated at first, since a major premise of his book was based on an assumption that simply wasn't true, and most scientists of the time suspected as much.[2]

We say that voltage is a potential for current to flow through a circuit that has resistance (and everything, even a wire, has some resistance). Think of it this way, your power

---

[1] We refer to "Amps" as a shorthand for the correct "Ampere." We use "I" because it stands for the *intensity* of the current (originated by the French). Also, we capitalize all three units, since they are named after the scientists that investigated and developed their concepts (they didn't invent them, of course).

[2] His assumption was that an electromotive force could only act when physical contact was made, which was contradicted by the Leyden jar, common at the time. He wasn't alone in history about not getting everything right—Newton believed in the transmutation processes of alchemy; Copernicus correctly removed the center of the universe from the Earth, but then placed it in the sun; and Einstein never accepted quantum mechanics— this is not to diminish their contributions, but merely to show that we are all immersed in the general assumptions—true or false—of our times.

utility applies voltage to your house, and delivers current (for which you pay) as you use it, i.e., as you insert resistances (plugging in your various appliances). Your electric meter measures the current flowing into your house.

We pause a moment. I just talked about current flowing into your house, as though the electrons arrive on one wire, and then leave on another after doing something useful in your house, like charging your phone. This would be exactly the case if the utility provided DC voltage, but, however useful it is to think of current flowing like water through your house doing useful things, the reality is that those electrons simply move back and forth (because it's AC). Further, in the time it takes for that alternating voltage to swing from positive to negative, and back to positive, all that those trillion upon trillion electrons do is essentially dance around in one place, but they're all dancing to the same tune, and that tune is, for example, ultimately loading your phone battery with a new charge.

So you're paying the power utility to make the electrons in your house dance back and forth in place.

I had to get that off my chest for the sake of complete disclosure, but I encourage you to go back to thinking about connecting a wire so that electricity can flow through it like water.

The alternating current in your house changes direction sixty times each second—the 60 cycle term you're probably familiar with. During the early development of power generation, there were different and isolated power stations for each local area, and each one ran at whatever frequency (cycle rate) was convenient. Since the operation of many motors is dependent on the powering frequency, the power stations had to get together and settle on one common rate. Westinghouse Electric, the largest power producer at the time, chose the 60 cycle value, since this was the lowest frequency that produced no noticeable flicker in the arc lamps that were prevalent at the time.

Once the power generation stations began connecting together to form what eventually became the current multination grid, each power generation station had to be precisely synchronous with all the rest. The turbines and generators of today's power plants are absolutely huge, and synchronizing all these together across the nation—multiple nations—is something of an engineering miracle in itself.

We take battery-operated wall clocks for granted now, but forty years ago, they were entirely dependent on the 60 cycles of their powering AC. As a fortunate side-effect of the power-generation synchronization, the 60 cycle rate is maintained at exactly that, and all the clocks had to do was click a gear along each power cycle. For every sixty clicks of that gear, the second hand advanced one second.

But, why the three plug contacts on your AC outlets? We know that two contacts are required to make a complete circuit, but what is that round contact on the bottom for? You may already know that it has something to do with safety, and here's how it works. The two flat blade contacts on the top of the outlet are the powering lines. The one on the right is the "hot" contact, and the one on the left is the "neutral" (assuming the third hole is at the bottom). The neutral line is tied to ground[3] back at the electrical panel, where the utility power meets your house. Normally, if you stick a knife in the left hole, you won't get shocked—please don't do this, though, since you could die, on the off-chance some yahoo has wired your house wrong. That round receptacle is also tied to ground. Here's the key part: manufacturers of appliances are careful to make sure that this round plug is connected to any conductive parts of the appliance that you might come into contact with—the obvious being a metal casing. The idea is that if something goes wrong in the appliance, and the "hot" line makes accidental contact with one of those places that you could touch, it would create a short circuit between the "hot" line and ground. This will cause the circuit breaker to trip and kill the "hot" line, saving your life in the process and

---

[3] It's called "ground," because at some point, the contact is actually made to the ground of the Earth, either via a stake, or the house plumbing, or even the metal casing of the utility feed.

producing howls of anger from those using their desktop computers on the same circuit.

Some appliances don't have that round third contact. Unless the device is an antique, one of the blade contacts is wider than the other, as you well know, since your first try at plugging it in is always backwards. This ensures that the neutral connection is made to the one the manufacturer intended. On these appliances, there is typically no external metal parts on which you might accidentally electrocute yourself. Instead, the keyed mating system ensures that the neutral line is connected to places you might contact with some effort, like the metal barrel of the screw bulb socket of a lamp. Wall adapters typically don't use a wider neutral blade, since you'd have to use a drill to get at anything dangerous, and this way you have more flexibility in the orientation when you plug it in.

Note that a circuit breaker doesn't save you if you inadvertently come into contact with 120 volts. Your body has a relatively high resistance, and would draw much less than an amp—too little to trip the breaker, but plenty enough to kill you.

But then there's those odd outlets in many bathrooms (all bathrooms in newer houses). They are designed to indeed save you if you manage to contact 120 volts. They include this extra measure of safety, since bathrooms mean water and plumbing, and water and plumbing means a quick and potentially deadly return path to ground if you contact the "hot" line. These are called GFCI outlets, and, not that you really care, that stands for "Ground Fault Circuit Interrupter." They are essentially AC outlets combined with a circuit breaker and a "residual current sensor." If you accidentally contact the "hot" line, by, for example, dropping the blow dryer in the bathtub, any unwelcomed current that goes through your body is sensed, and the outlet trips its internal breaker. Conveniently, there's a reset button provided to un-trip it, which is good, since they tend to trip more often than required. They are also used for outside sockets where contact with the (dangerous) earth is almost guaranteed.

An important point about these GFCI outlets—it is possible to use one GFCI outlet and then wire other outlets "downstream" so that all of them are protected by the one-and-only head-end GFCI. This is often done outside as a retrofit, where only the one head-end outlet is replaced. The result, though, is that you may find an outlet outside (or even inside) that inexplicably doesn't work, where you know it had previously. The problem may be that the head-end GFCI has tripped, and you'll have to go and hunt it down to reset it. While you're at it, you might also keep an eye out for the person who tripped it, and then fell backwards in surprise at the quick shock, hitting their head and knocking themselves unconscious.

So, if the main circuit breakers, located in the panel or box where the power utility delivers your feed, don't protect you from getting electrocuted, then what are they for? Primarily to prevent your house from burning down. The resistance of the wiring in your house is less than an ohm. If a direct short were to occur, 120 volts across one full ohm is, as we now know, over a hundred amps, so the short could easily result in two hundred amps or more; this is plenty to turn the weak point of the circuit into a red-hot heating element. The metal will eventually melt, breaking the circuit, but perhaps not before catching the paper on the insulation inside your wall on fire.

House circuits were originally protected with fuses, which are essentially dedicated weak points in the circuit, weak points that are guaranteed to be weaker than any other point, and where the metal can get red-hot and melt in a contained place. Long ago these were replaced by circuit breakers, which have the advantage of being reusable. Breakers for regular house circuits—the ones feeding the various AC outlets—are usually either 15 amps or 20 amps, depending on the expected loads, and subsequently the size of the wire used inside the walls. Higher amperage breakers—30, 40, or 50 amps—are usually dedicated to specific appliances that require more current. Examples are water heaters, clothes dryers, electric stove ranges, and hot tubs—devices that can inflate your electric bill if used extravagantly.

You might think that a 15 amp breaker will trip when anything over 15 amps is encountered, but that's not the case. The current rating defines how much current that breaker can handle on a continuous basis—it's the manufacturer's guarantee that it won't trip *below* that amount of current. So, when will the breaker trip? There's no fixed point. It's a matter of the amount of over-current, and time. The breaker may take a minute to finally trip at 22 amps, but in a couple of seconds at 90 amps, and some thousands of a second at 130 amps. This allows for surge currents—large, short-term current draws when devices, particularly motors, first turn on—but provides protection against flat-out dangerous shorts. This explains why you only get a few seconds of microwave time on your burrito when you're running the toaster oven, coffee maker, and waffle iron at the same time.

A reasonable question: how many appliances you can crowd together on the same circuit? It's obviously a matter of the total current. Some appliances list the current draw, but these are often the maximum peak current that this device could draw—the surge current we just talked about. A better number—better in the sense that it represents the average current draw—is derived from the power rating. This will require converting the power rating into current, and this is simply done. But first, we'll take a look at just what power means.

Back to the water analogy. Imagine a huge pipe with water flowing just an inch per minute. You could jump in and swim against it easily. This is like a lot of current at a very low voltage. Now imagine adding a high voltage—that same pipe feeding from the bottom of a dam. Not only could you not swim against it, but it would carry you away and the car you drove in on. The second scenario clearly represents more power, and you can see that the power is related to both the quantity of water, and the pressure behind it. In the same way, power in an electrical circuit is directly proportional to both the voltage and current. In fact, like Ohm's law, the relationship is simple arithmetic: electrical power, measured in watts, is the product of the voltage and the current. Thus, if I have 12 volts pushing 200 amps (e.g., I'm cranking my car's engine), I am delivering 2,400 watts. Alternatively, if I know I have 1,200 watts, and the voltage is 120 volts, then I know the current is 10 amps.

Note that the units are still volts and amps.

So, now we know how to determine how much current an appliance draws if we know its power rating. A 60 watt incandescent bulb draws 60/120 = 0.5 amps. A 1,200 watt vacuum cleaner draws 10 amps. We have to be careful with microwave ovens, since they advertise the amount of power they deliver to your food. Since they are not 100% efficient, their actual power draw is typically about 20% higher. So a "1,000 watt" microwave will also draw 10 amps.

Clearly, if I'm running my 1,200 watt microwave, and my 1,200 watt toaster oven and my 600 watt refrigerator all together on the same 20-amp kitchen circuit, I am drawing more than 20 amps. In fact, we can now calculate with a snap of the fingers how much—25 amps. The circuit breaker doesn't trip only because we generally don't run a microwave for more than a couple of minutes, and both the toaster oven and refrigerator usually cycle on and off as they maintain their internal temperatures. The reality is that a typical kitchen served by a 20 amp circuit is perennially skirting an outage.

Which is why most new homes are now provided with 30-amp circuits for the kitchen.

But what do you do when the display on the microwave suddenly goes blank, and the refrigerator ominously silent? You search out the breaker panel and look for one of the switches that's hovering in the middle, rather than all the way to the right—the "on" position. This is the breaker that has tripped. To un-trip it, you have to first push the switch all the way off (to the left), then back on (to the right, or sometimes down and then back up, if they're mounted sideways). If you hear a click, and the switch ends up back in the middle, it means you probably have a short somewhere, hopefully not a fork jammed into a wall socket.

One last word about circuit breakers. If you find that you've had to un-trip the same breaker a couple of times, you should check to see whether you're drawing too much current from your squadron of appliances. If that doesn't explain it, then the breaker itself may be dying, in which case it's time to call the electrician. Also, breakers are meant to be restored after a trip (otherwise we might as well stick with fuses), but they are not meant to be used like a light switch. Breaking the circuit is something of a dramatic event for them, and eventually they cave to the stress. Any tripped breaker means something needs to be addressed.

So far, we've talked about 120 volt circuits (sometimes referred to as 110 volts—the actual voltage is usually somewhere in between). You may have noticed, though, that some appliances—electric ranges, clothes dryers, and water heaters—require 240 volts. This is achieved by using two 120 volt feeds that are 180 degrees out of phase. This simply means that when one side is at its highest point in the 60-Hertz alternating cycle, the other side is at its lowest—one positive, and the other negative. Since they are both 120 volts away from neutral (ground), the voltage between them is the sum of the two.

But, how did we get these two lines that are 180 degrees out of phase in the first place? The utility company provides your house with two main feed lines—the out-of-phase lines we just described. Each one is 120 volts, and identical, except for being out of phase. The various 120 volt circuits in your house are approximately divided between these two. If you were to touch the hot wire of one circuit with one hand and a hot wire from another circuit with the other hand, you'd either feel nothing (because they happen to be sourced from the same feed), or you'd get knocked on your butt and maybe killed (because in this case, they are sourced from different main feeds, and your hands are bridging 240 volts).

Don't do this.

3

Heat and a Bit of Incidental Light

Toasters and Light Bulbs

Electricity has evolved from a scientific curiosity that drew early investigators outdoors into thunderstorms with kites, to the neural life-juice of your car that allows you to talk to it. It is the underpinning of our civilization: lifting our heavy loads, transporting us between cities, turning night into day, and animating the brains of our devices.

The multitude of services that electricity provides can be divided into three broad categories:

1) transform energy into other useful forms;

2) do work—move things;

3) control things and perform assigned tasks.

This chapter explores the first category, which sounds like something from a science fiction movie, but which is, ultimately, pretty mundane.

∞

Energy is everything. Sort of. Einstein told us that energy and mass are two forms of the same thing, but that's only relevant in the center of the sun where hydrogen is fused to form helium as a natural process, or the center of a man-made device where hydrogen is fused into helium for the purpose of annihilating cities. When I say that energy is everything, I mean that energy is the measure of everything.

Your airfare primarily covers the cost of aviation fuel, which is essentially solar energy that was stored via chlorophyll millions of years ago. It also covers the cost of the plane, which includes the materials—metal, which required energy to mine and extract from the ore, and plastics, which required energy to process from petroleum-based material, again the stored million-year-old solar energy. Your fare also pays the salaries of the pilots and attendants, for which they use the money to buy food (yet again, the result of solar energy), and housing (wood, metal, cement—all the result of using energy), and, of course, their electric bill, a very direct source of energy. Even the very cells of the pilot can be viewed as billions of molecules passing units of energy around. In a very real sense, our entire world is simply matter interacting at the impetus of energy.

Exactly what it is that we call energy, though, is not easy to put your finger on. You can't point to one thing, like a battery, and say, "That's it." A battery contains one form of energy, but it is *just* one form. Rolling a boulder up a hill seems fundamentally different from the light emitted from your flashlight, and one of the great landmarks of human enlightenment was understanding that they are not only related, but are manifestations of the same thing. If you know the quantity of one, you can exactly predict how it can be converted into the other. And when I say exactly, I mean within the resolution of our measuring abilities.

James Watt, an eighteenth-century Scottish engineer started the ball rolling (so to speak) by realizing, quantifying, and putting to practical use the idea that heat and work were related in a mathematically predicable way. Heat, of course, is just one form of energy, albeit a rather obvious one. James Watt did not originate the science of thermodynamics, but he took the first important step in that direction.[1]

---

[1] James Watt also did not invent the steam engine (the Newcomen engine had already existed for fifty years), but the improvements he made were so significant that the steam engine's then-sole function—that of extracting water from mines—was expanded to become the engines that powered trains and ships and launched the Industrial Revolution.

Thermodynamics is, according to the Concise Oxford Dictionary, "the branch of science concerned with the relations between heat and other forms of energy involved in physical and chemical processes." This is an appropriate segue back to electricity in your home. We've already explored what electrical power means in terms of voltage and current. Another way of defining power is the amount of energy consumed per unit time, i.e., the rate at which energy is being used. The more power, the more energy being used each second. There's an important distinction here. Power and energy are not equivalent. Power is the amount of energy being used at any moment. The amount of energy consumed in an hour is the average power level used over the course of that hour. Trotting out a metaphor, if power is the position of a faucet, then energy would be the number of gallons flowing per hour. We may call the utility company that supplies our electricity the "power company," but what they're really providing is energy in the form of voltage-pushed current. If one of your appliances is rated at 1,000 watts, then it's drawing a kilowatt of power. If you run that appliance for one hour, then you've used a kilowatt-hour of energy. You don't pay for kilowatts, you pay for kilowatt-hours. You pay for energy, not power.

Electricity, when delivered as both voltage and current, is energy. Energy is useful, particularly when provided in a convenient form. Like electricity. For hundreds of thousands of years, our critical energy source was wood, which we burned for heat and light. As we've seen, the early uses of electricity were for the same purpose, light being the primary component. Initially, they were one and the same, in the sense that Edison's light bulb (and all incandescent light bulbs) works by using electricity to heat a wire so hot that it glows and gives off light. In a sense, a light bulb is simply a resistor. A sixty watt bulb uses 0.5 amps (if power is "P," then $P = I*V$, or $I = P/V$, or $I = 60W/120V$). Ohm's law tells us that

---

Ironically, James Watt invented the horsepower as a unit of power measurement, which scientists have been trying to replace with the "watt" for over a century.

the resistance of the sixty watt bulb is (R=V/I) 120V/0.5 amps = 240 ohms.[2]

This may seem pretty low-tech, and it is compared to modern lighting sources, but it took Edison over a year of experimenting to find the right filament material (which ultimately was replaced by a better one, anyway). But the filament material wasn't even the high-tech (for the time) aspect. If you were to break the glass of an incandescent bulb and turn it on, it would glow and then dramatically disintegrate in a matter of seconds. Exposed to the oxygen in the air, it burns—even tungsten, which is metal. Light bulbs must either be evacuated (create a vacuum), or filled with an inert gas (e.g., argon or krypton). This was the high-tech aspect.

Let's take a closer look at how electricity creates heat. I said that the filament of an incandescent bulb is effectively just a resistor. Do all resistors get hot? No, but "hot" is relative—all resistors *do* give off heat. It's just that the resistance of an incandescent bulb gives off a *lot* of heat. The reason that all resistors give off heat is that they are dissipating the energy delivered by the voltage and current. The wheel hubs of your car get warm after a lot of stop-and-go driving. This is because of the friction created between the pads and the brake disk (or drum). The energy of your moving car was converted into heat in your brakes. In the same sense, the energy of the moving electrons are converted into heat in the resistor as it "resists" the flow.

In a resistor, the heat is entirely dependent on how much current is flowing through it. You might think that twice as much current would create twice as much heat (i.e., would dissipate twice as much energy), but this is not the case. The heat created is not in a linear relationship with the current. Twice as much current creates four times as much heat.

---

[2] If you own a multimeter and ran off and measured the resistance of one of your sixty watt bulbs, you should have read this first. The resistance of the tungsten filament of an incandescent bulb increases as it heats up. When cold, the resistance is usually less than twenty Ohms. This is why bulbs usually blow when you first turn them on—there's a large inrush of current until the filament can get hot.

Ohm's law tells us that in order to push twice as much current through our resistor, we would have to double the voltage. We know that power (the instantaneous quantity of energy) is voltage times current. We have just doubled both of them. Two times a doubling is times-four.

As it turns out, creating heat is the most efficient use of electricity, when efficiency is the measure of how much of the energy delivered is converted into the result we want. The reason falls back to thermodynamics—the second of the four laws, which, in fact, when applied to everyday experiences, essentially states that you can't do anything (i.e., use energy in some fashion) without creating some amount of heat as a byproduct.[3] We've seen how stopping your car generates heat in the brake pads, but heat is also generated, wanted or not, by vacuuming the carpet, using a fan, rubbing out a mistake with an erasure, or by your computer or phone just going about its business doing nothing but toggling invisible bits. Most of the time, the extra heat is unwanted, wasted money paid to the utility company, or reduced time on a rechargeable battery. But if its heat you're after in the first place, then anything wasted is thrown back into the pot.

Creating heat with electricity may be its most efficient use, but that's viewing it from the abstract perspective of physics. It doesn't necessarily mean that it's the most cost effective use, not by a long shot. The reason is that there's lots of ways to generate heat (again, thanks to the second law of thermodynamics)—ways that aren't good for much else *but* generating heat. Natural gas can be used in absorption-type refrigerators, but, other than recreational vehicles, you'll never see one. If natural gas is provided to your home, it is used for heating your house, heating your water (via the water heater), and possibly heating your clothes (via the clothes dryer). If you live in a cabin in the woods, you might burn wood for heat. Good luck trying to use propane gas or

---

[3] Looked at in a more general sense, the second law states that processes in nature—i.e., using energy in some way—is simply always going to be inefficient to some degree. This is why there is no such thing as a perpetual motion machine. Yet another way of stating it is that entropy (the degree of un-orderliness) always increases overall.

split pine to power your widescreen TV.[4]

The problem is that the amount of heat needed to warm a whole house, or even a room, is huge compared to, for example, the heat needed to cook your single-serving dinner in the microwave. The bottom line is that we have much more useful things to do with electricity than to (virtually) throw it away as heat. After all, unless you live near a hydroelectric dam, your electricity was generated using heat (burning either gas or coal, or catching speedy neutrons in a nuclear reactor). The utility company went to great expense (your expense) to convert that heat energy into a versatile, broadly usable form. Heating your whole home with electricity is like shoving a diamond engagement ring under the leg of a table to stabilize it.

That said, using electricity to generate heat makes a lot of sense when applied to specific, localized needs. It would be cumbersome indeed to run a natural gas line to your coffeemaker. In these cases—coffeemakers, toasters, clothes irons—using electricity as heat is more like using the diamond engagement ring band to rub a scratch-off lottery ticket.

The basic operation of almost any appliance whose primary function is to use heat hasn't changed for many decades. They all push current through a resistance element in the form of, for example, a heating coil. Besides coffeemakers, toasters, and clothes irons, these categories include hair curlers, electric blankets, blow dryers, heating pads, space heaters, and waffle makers. Heat is the most basic form of energy, but we put it to a lot of uses. Obviously improvements have been made over the years, but they consist almost exclusively in automatically controlling when the heating element is turned on. Once turned on, the element works the same as it did a hundred years ago. Manufacturers often try to fool you into thinking they have a more efficient product, but they all convert 1 watt-hour of

---

[4] We're not talking about propane-powered electric generators—that would be cheating.

electricity into .293 BTUs of heat. Otherwise, they'd be violating a basic law of nature. The best they can do is not produce more heat than required for the job.

They can also avoid burning your house down. Many appliances have overheat protection, whereby a temperature-sensitive element breaks the circuit above a certain limit. These are common in types of appliances susceptible to inadvertent misuse. Common candidates are blow dryers that would eventually melt if the air intake is blocked—your own little personal nuclear reactor meltdown. Coffee makers are another common application. If you leave the pot on the burner until all the coffee evaporates away, the glass pot and gooey coffee residue insulate the burner, again resulting in a mini catastrophic disintegration if not circumvented with the overheat cutoff. Modern electronically controlled versions may require that you turn the appliance completely off if tripped—sort of a little penance ritual to make you more careful next time. But even the old-fashioned dumb ones require the temperature to fall below the original kick-out point before re-activating. This is partly an additional safety feature, but more importantly, it avoids the sputtering effect, where the appliance turns on and off repeatedly in short bursts. Besides being annoying, it can be interpreted by people who haven't read this book as a faulty unit, and the manufacturer would much prefer the fault be recognized as the user and not them.

In engineering, this type of on/off operation is called hysteresis. It applies to many different processes, not just heating elements. It simply means that any time the process includes an action triggered by a threshold being crossed by a rising level (e.g., the heating element being turned off), if the process incorporates hysteresis, then the undoing of the action (e.g., turning the heating element back on) occurs at a lower threshold. In other words, the threshold for a rising level is higher than that for a falling level. Once tripped, you have to back up a ways to un-trip.

We've seen how using electricity as heat in small appliances makes sense, while using it to heat an entire house is wasteful. An electric stove lies somewhere in between. It's a lot of valuable electricity turned into mundane heat, but there's just no practical alternative when your kitchen is not plumbed with natural gas (or propane, for remote areas). Here, the heating element is quite visible—you set your pans on it. In fact, that's not quite correct. The actual resistive heating wire —nichrome—is buried inside a strong steel casing. This is what you place your pans on. Additionally, the inner heating element is also surrounded by a ceramic casing. This provides electrical isolation so you don't get a 120V shock (240 volts, if you were able to touch the two ends of the heating wire coil—one in each hand. Don't do this).

In newer stoves, the heating coils may lie beneath a smooth, flat glass-ceramic top, providing an easier surface to clean, but vulnerable to scratching. Additionally, whereas the traditional heating coils convey heat by conduction (the heat is transferred directly by contact between the coil and pan), the glass-ceramic stoves heat via infra-red radiation—radiant heat. This means that once the stove is turned off, the glass-ceramic surface cools more quickly, and is less prone to cause accidental burns.

Older stoves provided a limited range of cooking temperatures by activating different, separate coil elements, or in combinations. Newer coil-type stoves control the cooking temperature by adjusting the average amount of current fed to the single coil. This is done via time averaging. Current is provided to the coil in bursts. When the temperature is turned to low, the current is provided in short bursts. As the knob is turned up, the bursts get longer and longer, until the temperature knob is turned all the way up, when the current is fed continuously. You're probably wondering why this is not done by adjusting the voltage, and I'm glad you are (wondering), since it shows you've been paying attention. Adjusting the voltage would indeed provide control over the current through the coil (Ohm's law), and consequently the heat generated. However, when dealing with the amount of current used in a stove, trying to do anything other than making and breaking the connection is not easy. It can be done for sure, but is more expensive.[5]

∞

We've been describing the most basic use of electricity, where its energy is simply transformed into another form. So far, the only form we've discussed is heat. Another form is light. To be accurate, the visible light from a floor lamp is actually the same form of energy as the radiant heat from a space heater.[6] Both are electromagnetic energy, they just wiggle at different frequencies—visible light at about five hundred trillion times a second, and infrared at only one hundred billion. From a practical point of view, however, we think of them as different forms, since they are used differently—one to keep from shivering, and the other to keep from walking into furniture.

As we've already noted in chapter 1, the first electric lights available in homes, and still popular until fairly recently, were incandescent bulbs. Incandescence means simply to glow with heat. A principle of physics is that the higher the temperature of an object, the higher the frequency of light it gives off, i.e., the more times it wiggles each second. A space heater is technically a type of incandescent light, but it's heated to a temperature where most of the light is infrared, which is what radiant heat primarily consists of. Some of the light inches up into the visible range, making the heating elements glow red. An incandescent light bulb is simply heated to an even higher temperature, where the emitted light spans the visible range. In a space heater, 99% of the energy is converted into infrared radiant heat, and 1% emitted as visible red light. In a light bulb, 98% of the energy is emitted as radiant heat, and 2% as visible light. An

---

[5] For decades, in fact, the method has been decidedly low-tech. A metal strip in the switch behind the control knob heats up, bends, and breaks a connection to cut off the current to the coil. The metal strip then cools off, un-bends, and makes the connection again, starting the cycle all over. The control knob adjusts how much the metal strip has to bend to break the connection.

[6] But not the warm air provided via a built-in fan. That's heat conducted from the heating elements directly to the air, and also indirectly some of the radiated heat that wiggled the air molecules.

incandescent light bulb is essentially a small space heater that incidentally gives off some visible light.

With only 2% of the energy delivered in the desired form, you can understand the impetus to replace the incandescent bulb with more efficient technologies.[7]

The earliest contender, and one that has been with us all of our lives, is, of course, the fluorescent bulb. The arrays of long tubes have become iconically associated with commercial applications, e.g., offices and retail stores. In fact, for decades, they've been ubiquitous outside the home. The reason is that, whereas a 60 watt incandescent bulb is 2% efficient, an equivalent fluorescent lamp is 25% efficient—a twelve-fold improvement (the actual practical improvement in efficiency is more like six-fold for reasons you probably don't care about). The obvious question is why the incandescent bulb lasted as long as it did. The answer is convenience and esthetics.

To understand our century-long preference for a white-hot tungsten filament, we first need to look at how a fluorescent light works. The operation and construction of a fluorescent bulb is significantly more sophisticated than that of an incandescent one. If an incandescent bulb is a touchtone landline phone, then a fluorescent bulb is a smart phone.

The difficulty in manufacturing an incandescent bulb is evacuating the air. In producing a fluorescent bulb, this is just the beginning. Once completely evacuated, the fluorescent bulb is filled with a mixture of an inert gas and mercury vapor. In order for the mercury to become a gas, and not remain the shiny liquid used in early thermostats (and now banned as highly toxic), the pressure inside the tube must be kept below 1% of atmospheric pressure. A tiny amount of mercury is mixed with an inert gas (e.g., argon) and electrodes are positioned at opposite ends of the tube. When the powering voltage is first applied (120 volts in the USA), a special circuit heats the electrodes, causing electrons

---

[7] It should be noted that some halogen bulbs can reach 3.5% efficiency. Yahoo.

to be driven off via the same process used in the old tube radios and televisions[8] called thermionic emission. These electrons bump drunkenly into the mercury gas molecules, causing them enough irritation to toss off a photon. Photons, are, of course, light, and the ones tossed off by the mercury atoms are too energetic to be seen. They are in the ultraviolet area of the spectrum. The inside of the tube is coated with a substance that absorbs the ultraviolet photons from the mercury gas, chews on it a bit, and spits another one back out, but this time in the visible range that we can see. This process of absorbing and re-emitting a photon (light) is called fluorescence.[9]

The argon gas acts as a sort of moderator, forcing the thermionic electrons to dance around and have a better chance of whacking into a mercury atom.

As the tube gets into gear, the effective resistance across it falls dramatically. In fact, the more current that flows, the more the resistance falls, allowing even more current to flow—you see where this is going. In order to prevent the tube from dramatically self-destructing, this falling resistance (called negative differential resistance) must be countered. This is the role of the ballast—those rectangular lumps hiding away above the tubes in the fixture. These are the sources of the humming that can develop as the fixture ages.

Because of their large fixtures, harsh light, and reputation for flickering and humming, traditional fluorescent lighting found limited use in the average home—relegated to the garage, and perhaps the kitchen. In response to the growing appreciation of efficiency, though, engineers addressed these problems and shrunk the form down until they could be used as replacements in the standard

---

[8] And still used in many guitar amplifiers. When driven past their standard operating parameters, tubes "mis-operate," i.e., distort, in a characteristic fashion beloved by rock and blues guitarists. Of course, since the tubes in these amplifiers are routinely driven into these distortion areas, that has now become the standard operating area.

[9] The difference between fluorescence and phosphorescence is that the former chews on the absorbed photon for a tiny fraction of a second, while phosphorescent substances can chew for hours. Glow-in-the-dark clock dials are an example of phosphorescence.

incandescent screw socket. These CFLs (compact fluorescent lamp) were impossible before modern electronics were able to replace the klunky humming black box ballasts.

By the way, the basic operation of the fluorescent light was demonstrated in the late nineteenth century, but it was essentially magic until the new science of quantum mechanics explained what was going on.

Also by the way, an incandescent filament reacts relatively slowly to changes in current, but a fluorescent bulb reacts quickly enough to shut down between the 60 cycle peaks. Your eye luckily averages it out, but some people can notice the flicker. Flies brains are tiny, but quick, and their world under a fluorescent light is one of a constantly strobing sequence of light and dark.

Finally by the way, neon lights are just a different form of fluorescent bulb. I should say that fluorescent bulbs are a different form of neon light, since the neon light came first.

∞

Before leaving the subject of lighting, we'll review two other types, the oldest, and the newest.

Arc lamps pre-dated Edison's incandescent bulb by more than fifty years. In fact, before 1900, arc lamps constituted the primary practical use of electricity (the telegraph aside). Because Thomas Edison advanced electrical generation in conjunction with his incandescent bulb, many people first experienced electric light in that form, and associated his bulb with the introduction of electric lighting.

Arc lamp operation is about as basic as you can get. If you've ever touched two flashlight battery wires together and seen a spark (or car battery wires together and shrieked), you've seen an arc light. I should explain that "arc lamp" here is specifically the carbon arc light, the most common type for nearly two hundred years, and the type traditionally associated with the term—fluorescent lights are also, technically, a type of arc light.

Arc lamps are started by briefly touching the tips of opposing carbon rods together, and then slowly pulling them apart. The initial spark—the same type you see when touching flashlight battery wires—heats the air, ionizing it, which means simply that the air molecules are so hot, one or more electrons are goosed free. Since electricity consists of moving electrons, these freely roaming ones cause the ionized air to easily conduct current. In other words, the spark produces a bit of air with a low resistance, allowing current to continue to flow, even though the carbon rods are no longer in contact. As long as this "arc" of ionized air is not broken, the current flows across the virtual wire.

The ionized arc is hot—thousands of degrees—and the tips of the carbon rods vaporize, slowly disintegrating. This is a desired part of the operation, as it's the hot carbon vapor that produces the light. The light of an arc lamp is intense and harsh, and after the widespread acceptance of the incandescent bulb, was used mainly to illuminate large areas, like arenas and factories.

As it happens, an arc light is very high in ultra-violet, rivaling that of the sun, and the carbon arc lamp continued for awhile to be found in some homes as a specialty health product—a substitute for natural sunlight. So, the tanning sunlamp goes all the way back to the roaring twenties.

The long-enduring carbon arc lamp has now been superseded by the xenon arc lamp—where the vaporized carbon is replaced with xenon gas. Like the carbon arc lamp, the xenon version also closely mimics the natural light of the sun.

The final type of light is the result of advances in semiconductor science. These are LED lamps, of course. LEDs (light-emitting diodes) have been around from the beginning of the digital revolution, starting in the seventies, but until recently, were not powerful enough—meaning the amount of current they could handle, and thus the consequent light—for general illumination, limiting them mostly to device displays. If you're old enough to remember the nineties, you'll remember that LEDs were limited to red, green, and yellow. In addition to developing LEDs that were powerful, the invention of the blue LED in the mid-nineties (commercially available after 2000) was necessary before they could be used for illumination. This was because we don't generally like to light our houses with red, green, or yellow light, or any combination of these. Blue was necessary to finally produce the white light we're used to.[10]

One of the main reasons to use LEDs for lighting is, of course, that they are three times more efficient than CFLs, and sixteen times more efficient than incandescent bulbs. But besides being efficient, LEDs have long lives. Very long lives. An LED lamp that's used four hours a day on average should last at least fifty years, although you might notice some dimming the last couple of decades (assuming you could remember how bright the lamp was thirty years before).

As their name indicates, LEDs are diodes that happen to emit light. We will explore this in more detail later, but for now, we note that diodes are polarized, i.e., they only conduct current in one direction (this is why you have to insert the battery holder in the flashlight the right way). Also, when sufficient voltage is applied in that proper direction, they have effectively no resistance, so we have to limit the current in some way, or they burn out. In applications where cell batteries are used, such as AA, AAA, etc., a simple resister suffices. Lithium coin cells, on the other hand, can often drive the LED directly. Diodes, including LEDs, require anywhere from 0.7 volts to 3.6 volts (more for the newest super-large LEDs) to conduct current. This is convenient when operating from combinations of 1.5 volt cell batteries, but problematic when used with 120 volt AC. LED lamps that are powered from an AC outlet include semiconductor electronics to convert the high-voltage AC to lower voltage DC required by the LED diodes, and provide appropriate current limiting.

So, long before the LEDs themselves give out, some component of the AC conversion electronics will likely let the team down.

―――――――――――――――――――

[10] Although white light can be achieved by using a combination of red, green, and blue LEDs (the familiar RGB combination), in practice, white LED light is produced exclusively with powerful blue LEDs that are coated with a fluorescent phosphor that emits yellow light, and the combination of blue and yellow appears white. So when all is said and done, LEDs lamps could be thought of as, at least partially, fluorescent lights.

∞

To many of us, even the CFLs are still new technology. For all of our lives, a 60-watt light bulb was a standard unit of lighting. If you wanted something more sublime, you went with 40 watts, if you needed to illuminate a large room, you picked up a 100-watt bulb. We equated the power usage with the light output, which was wrong, since, as we saw, an incandescent bulb is only 2% efficient. It would have been more accurate to think of a 60-watt bulb as providing 1.2 watts of light, which, obviously, nobody did.

As the more efficient technologies entered the market, the manufacturers needed some way to express the amount of light the product would provide. The easiest method was to compare the light to that of an equivalent incandescent bulb. Thus, a 13-watt CFL was advertised as equivalent to a 60-watt incandescent bulb; a 9-watt LED bulb, the same. This works fine for awhile, but there's a danger that we're creating a new, artificial unit of measurement. We already struggle with feet and ounces against the rest of the world's meters and grams, the last thing we want is another horsepower (versus the sensible watt).

Science, our friend, stepped up to the rescue. The lumen is a unit of measurement for light. Like the watt, it is a measure of instantaneous energy, but specifically the energy contained in light.[11] A typical 40 watt incandescent light bulb will provide about 520 lumens. A 60 watt bulb, 800 lumens, and a 100 watt bulb, 1300 lumens.

You may have run across another unit of light measurement called the candela, and there's an important difference between the two. Whereas the lumen is a measure of the total output of a light source, the candela is a measure of the amount of light passing through a specific area (a radial surface at right angles to the source), so candelas are a

---

[11] Specifically, the energy contained in wavelengths of light visible to the human eye. Because so much of the light energy given off by an arc light is ultraviolet, a lumens measurement is not a very accurate indication of the total electromagnetic output. Lumens, though, are still useful in comparing the useable light of an arc lamp compared to, say, a halogen lamp.

measure of the light intensity in *one direction*. Think of it this way: a typical triple-AAA battery LED flashlight might provide a total of 200 lumens. This is the same amount of (total) light that a 15 watt incandescent light bulb would produce—not much more than a nightlight. But a flashlight focuses all the light into a relatively tight beam, so the intensity is much greater. The candela measurement of the flashlight at, say, five feet, would be greater than that of a 100 watt bulb at the same distance.

Watts, BTUs, and lumens—all measures of static abstractions. It's time to get things moving.

4

Move Things

Vacuum Cleaners and Doorbells

In the previous chapter, I divided the general uses of electricity into three categories:

1) transform energy into other useful forms;
2) do work—move things;
3) control things and perform assigned tasks.

In this chapter, we'll explore how electricity is used to move things. The things moved range from air (fans) to garage doors, from dirt (vacuum cleaner) to smaller and smaller pieces of a peach (food processor). This chapter could technically be considered a subset of the previous one, since moving things involves work, and work is just another form of energy. In physics, the definition of work is energy transferred by a force acting through a distance. The two key words here are "force" and "distance." Work involves both applying a force to something, and then that something moving some distance. Your boss may tell you to push against a brick wall—i.e., apply a force—all day long, but when you go home some hours later exhausted and hungry, you haven't done any work, at least by the definition in physics, since the wall hasn't moved.

A more general definition of work is found in the field of thermodynamics as the amount of energy transferred from one system to another that isn't dependent on heat. Note that we're not saying that there *is* no heat—no system is perfect, and thus *some* amount of heat is always generated—but simply that heat isn't used as a vehicle for energy transfer. The energy transferred always results in something being moved, though, whether it's a one-ton rock by a bulldozer, or atoms from one molecule to another as in chemical reactions.

Doing work is moving things, but how does electricity, this invisible, weightless energy living in the wires behind your walls, manage to move physical objects? It comes down to magnetism. If you hold a magnet close to your refrigerator, it will make a little jump to grab hold. Magnetism moved the magnet. If you hold the magnet over a paperclip, the paperclip jumps to the magnet. Magnetism moved the paperclip. Put generally, magnetism is a force that works over a distance—as long as it's working with either iron or nickel, or another magnet.

Here's where we pull the rabbit out of the hat. It turns out[1] that if electricity is passed through a wire, a magnetic field forms in a circle around the wire. In the everyday world, this magnetic field is pretty weak; unless you're passing hundreds of amps through the wire (a really thick wire), you'll need a special tool to even detect it.

But if we make a loop in the wire, the circular magnetic field concentrates in the center (just as it spreads out around the edge). Imagine buying a sausage wrapped in plastic, and bending it into a tight circle. The plastic bunches together in the center, like the concentrated magnetic field. If, instead of just one loop, we make two loops and place them side-by-side, like would happen if you turned the wire twice around a pencil, now the magnetic fields of the two loops overlap and add together, doubling the strength inside the circle. We can

---

[1] It doesn't just "turn out" of course. Electromagnetism is predicted by Einstein's special theory of relativity, and can be explained using quantum physics. For you and I, though, it just turns out.

add more turns, but as they spread out across the length of the pencil, only adjacent loops overlap, and the concentrated magnetic field starts to smear across the length of the pencil. So, instead of continuing all the way down the pencil, after, say, ten turns, we reverse directions and starting winding the turns back, on top of the first row. The magnetic field of the new (overlaid) row of turns aligns with and overlaps that of the first (buried) row, strengthening the density of the field in the middle of the pencil. Each overlaid row adds greater strength.

But we still have this concentration of magnetic field smeared along the length of the center of the pencil. Please welcome iron to the show. Iron has a property called ferromagnetism, which means that when iron encounters a magnetic field, its atoms respond by aligning themselves to the field. They become, in effect, a conduit for the field. If we replace the pencil with an iron (or nickel) bar, all that magnetic strength that's been concentrated along the length now becomes available at the ends. The magnetic field strength at the end of the bar is essentially as strong as that of all the turns added together. This can be substantial. If we place a thousand turns around the bar, this "electromagnet" becomes a thousand times as strong (approximately) as any single turn.[2]

Like any magnet, an electromagnet has poles. One end of the bar is north, and the other is south. Which is which depends on the direction of the current flowing in the wire. If we reverse the direction of the current, the poles swap places. (This will be important later.)

Another property of electromagnetism is that the strength of the magnetic field surrounding a wire is proportional to the amount of current flowing through it. Double the current through our coiled electromagnet, and it can pick up a piece of iron (or another electromagnet) that's

---

[2] There are limits. Try to force too much magnetic force through the iron bar, and it will saturate, meaning that it just can't handle any more. In this case, all the atoms are now aligned.

twice as heavy. Add enough turns of wire, and current—and greater mass of iron for a core—and the electromagnet can pick up a car—as you can observe at an auto-wrecking facility.

All this is fine in theory, but there are not a lot of uses in your home for the ability to grab and hang onto a hunk of iron. Actually, if the hunk of iron is the keys to your car, and you have a teenager living at home, this could indeed be a useful feature. But that aside, what we really want, obviously, is a motor.

We already have a mechanism to pull on a piece of iron or magnet, we just need it to go around and around in a circle. Imagine that you're standing on a rotatable plate, a Lazy Susan, say. Imagine also that your arms are outstretched and you're holding a red ball in one hand, and a blue ball in the other. Now add two friends on either side of you. One of them hates red, but loves blue, and the other is the opposite—she hates blue, but loves red. The blue-lover reaches out and pulls on the hand with the blue ball, while pushing away the hated red ball. The other friend, meanwhile, pulls on the red ball, and pushes on the blue ball. Your friends are working together to rotate you. But, just as your red ball is about to align with the red-lover, and the blue ball with the blue-lover, you trick them, and switch the balls. Now, the red-lover suddenly sees the hated blue ball going by, and gives it a shove away, and the other friend does the same to the red ball. If you are adept at switching the balls at just the right time, your friends will spin you around and around, at least until you become dizzy and drop the balls.

You probably see the operation we're getting at. Replace your friends with magnets, one facing its north pole inwards, and the other its south pole. Your red and blue balls become coils of wire around a piece of iron facing outward—two electromagnets. These two electromagnets always have current flowing in opposite directions, so when the outward-facing pole of one is north, the other is south. So far we've recreated your friends and you with your red and blue balls. All we need now is a mechanism to switch the balls at the right time. This is the tricky part. We replaced your red and blue balls with coils of wires. But you represent the rotating shaft of the motor, and we have to somehow get electricity to your coils. What we'll do is position four curved electrical contact plates in a circle around the motor shaft (say, around your ankles), and connect opposite pairs to the each of the wire coils. Then we'll touch two pairs of fixed contacts against the shaft that provide two sources of electricity, each of an opposite polarity. We position the shaft contacts so that as we turn the shaft, each coil is sourced from one set of contacts for half a turn, and then the other for the other half of the turn. So we have coils (electromagnets) whose direction of current (and magnetic polarity) switches each shaft rotation.

Putting names to the parts: you (the rotating shaft) are called the rotor; the balls (the iron cores of the coils) are the armature; your friends (the fixed magnets) are part of the stator; your ankles (where the curved contacts are positioned) are the commutator; and the external contacts are the brushes.

As you might guess, this represents a very basic motor. Varying the amount of current flowing through the rotor coils changes the "pull" of the electromagnetic field, and consequently the strength of the motor. This can control the speed of the motor, but only if the load is constant. Actual motors used in appliances will usually have multiple coils and more complicated commutators to smooth the operation and reduce wear on the brushes. Powerful motors, as you might find in your washing machine, may use electromagnets in place of the fixed magnets.

When more precise control over the motor speed is needed, as in computer disk drives, the fixed magnets can be placed on the rotor, and the current-reversing electromagnets on the stator. Now there is no physical commutator or brushes. This is called a brushless DC motor. Note, however, that the brushes/commutator design is self-aligning, meaning the rotor coils change polarity automatically at the right time as the shaft turns. When we remove the commutator, we need an alternate method of changing the coil polarity. This involves a sensor system that tracks the shaft's position, and is obviously a whole level of additional sophistication (which, thankfully for you, we won't go into).

∞

Working motors were developed early in the nineteenth century, but their general application had to wait until power generation and distribution was introduced in the 1890s. Once they became available, though, they launched the second phase of the industrial revolution. Compared to steam engines, electric motors are compact and flexible. Mechanical power could be placed exactly where it was needed, eliminating the belts, pulleys, and chains of steam-powered industry, vastly reducing maintenance. Also, with power generation centralized (and sold as a product), factories could focus on just the machinery required for their operation, eliminating the redundant steam engines owned, operated, and maintained by each factory—in many cases, multiple steam engines.

Early motors were expensive, though, and although a boon to factory operation, due to their high cost, motor-based home appliances were a slow start, initially reserved for the well-to-do. It was the newly-motorized factories themselves that finally automated motor manufacture to the point that the average home owner could afford them. Today, over half of all the electricity in the United States is used by motors.

Modern electric motors have one significant limitation, however—they are rotational machines. This is great for appliances that want exactly that, like fans, clocks, and food processors, but not so convenient—back to the gears or belts—for devices like garage door openers, clothes dryers, and DVD player disk trays. Some appliances may seem as though they would be amenable to a rotational power source, but in fact require some fudging to make do. One example is a food mixer. The mixing blades do indeed rotate, but there are two of them. Unless you want to try to use two synchronized motors, you're stuck with some gearing. Another, much more complicated, example is a top-loading washing machine. The agitator must turn back and forth repeatedly (during the wash cycle), and also spin quickly (during the drain/spin cycle). The mechanical design of all this clutching and cranking represents a surprising degree of sophistication, and it's only because the machines have had half a century to mature that they can operate for years without failing. A front-loading washing machine uses a simpler mechanical design (comparable to a clothes dryer), but even here, the motor is connected to the drum via a belt.

∞

Every home has one appliance whose core operation uses a large motor, but whose visible function has no moving parts.[3] I'm talking, of course, about your refrigerator. There are other ways to "make coldness," but using the thermodynamic principles of the combined gas laws is far and above the most ubiquitous. A refrigerator is essentially a heat pump. Let me correct that: a refrigerator is exactly a heat pump (as is an air conditioner), but unlike the home heating/cooling units that were all the fad in the 1980s, a refrigerator's heat cycle goes in just one direction (pumping heat from inside to outside the box—into your home). In a nutshell, a thermodynamically appropriate gas (with a chemical name so long that they use abbreviations—formerly Freon until we discovered it was yanking off the Earth's ozone blanket) is pumped by the motor-driven compressor through some pipes. Seriously, that's it. There's nothing else inside the nutshell.

The cooling trick is manifested in the shapes of the pipes. The key property of all gasses is that when forced into a smaller volume—into a smaller diameter pipe, for example— the temperature goes up. The energy content is the same, but compressing it down into a smaller size manifests in a higher temperature. The higher temperature makes it easier to

---

[3] The absolute scope of "every home" may seem like a bold and risky statement, but I'll bet you don't know of a home without one. We're not talking about a homeless person's cardboard box, since the definition of "homeless" is having no home. In fact, I'll offer this precise definition of homelessness: a domicile with no operable refrigerator. In the 1930s, the first appliance most homeowners bought was either a refrigerator (if you were affluent, and had servants to do the cleaning), or a vacuum cleaner (because it was cheaper).

dump the heat into the air, and this is helped along by running the hotter, compressed gas through a long, winding course of squiggly turns that covers most of the back of your refrigerator. All the heat—the energy—removed from inside your refrigerator in order to make it cold, is shoved out behind it. As the heat is dissipated into your kitchen, and the temperature of the gas in the squiggly tubes falls, it condenses into a liquid. Thus, this part of the tubular journey is called the condenser. Most refrigerators are pushed back against a wall in the kitchen, and the manufacturer installs standoffs in the corners to prevent you from pushing the fridge back so far that the condenser coils contact the wall and block airflow. Otherwise, the heat wouldn't be able to escape, and the compressor motor would fail as it exhausted itself in the effort. Some models use a fan to blow air across a more compact cluster of condenser coils. This allows for more flexible design—and another component to fail.

By the end of the condenser portion of the cycle, the gas is liquid, cooled to near-room temperature. Think of this as a sponge that we've wrung dry. We've gotten rid of the heat from the previous cycle, and we're ready to extract more from the interior of the refrigerator. We're ready to sop up more heat with our gas sponge. We do this by allowing the room-temperature liquid to expand back into a gas. Just as a compressed gas goes up in temperature, an expanded gas goes down. As the liquid gas expands and reforms into gas, the temperature drops inside the freezer. Heat from the freezer is absorbed by the expanded gas, raising its temperature, and lowering the temperature of the freezer—they meet somewhere in between. Because the liquid gas was allowed to expand back into a gas, this part of the cycle is called the evaporator.

We've now completed the refrigeration cycle. The sponge is again full of heat, ready to be compressed in the condenser and dissipated from behind the fridge.

The evaporator is located in the freezer compartment, and the refrigerator area is kept cold via a fan that blows in cold air from the freezer. The "coldness" dial for the refrigerator sets a thermostat that controls the refrigerator compartment temperature by both turning on and off this circulation fan, and simultaneously running the refrigeration cycle for awhile. The temperature of the freezer, therefore, is dependent on the setting of the refrigerator. The separate freezer dial adjusts how much air is circulated between the freezer and refrigerator compartment, thus indirectly adjusting the average temperature of the freezer. This scheme establishes a fairly precise temperature in the refrigerator compartment, where, for example, we want to keep the milk as cold as possible without actually freezing it. We don't care so much about the exact temperature in the freezer, so long as everything stays frozen.

Perhaps the greatest advancement in the engineering of refrigerators has been their reliability. This is almost entirely due to the fact the electric motor is sealed along with the compressor mechanism inside the pipes that circulate the refrigerant gas. Once assembled, all the joints in the system are soldered, the gas loaded, and the final fill-hole permanently plugged. The only things that go in or out of the "closed" system are electricity and heat. This is why when you buy a new refrigerator, you plug it in, and don't think further about it for ten, twenty, or even thirty years. Try this with a steam engine. This is also why, on the other hand, when the electric motor does eventually fail, you either call a special refrigeration technician, or, more likely, buy a new refrigerator.

This degree of reliability was not always the case in the early years, with dire consequences. Early refrigerators used gases that were toxic, causing blindness or death if inhaled. Failure wasn't common, but catastrophic when it happened. The fact that people continued to buy these and gamble with the risk was a testament to the degree of life improvements they afforded. Imagine today giving your daughter a smart device, knowing that over the next year, someone in your county was going to die from using one.

Although potentially deadly, many of the early refrigerators were a lot more efficient than today's versions.

This is because we're willing to pay for the convenience of automatic defrosting. Every time you open the door, you let in a new batch of moisture which tends to condense on the coldest surface, which is, of course, the evaporator coils, where it freezes. Besides eventually commandeering all the space in the freezer, the ice insulates the coils, making it harder and harder for the compressor to keep the rest cold. The older models, and still today with virtually all the mini units, the entire refrigerator had to be periodically powered down, and the ice allowed to melt. The "frost free" models achieve automatic defrosting by every now and then shutting down the refrigeration cycle and activating heating elements, which melt off any frost that has accumulated. This obviously runs counter to the goal of keeping everything inside as cold as possible, and detracts from the long-term, average efficiency. Of course, shutting down, warming up, and then re-cooling the non-automatic units wasn't exactly efficient either.

By the way, the reason today for the different efficiency ratings among refrigerators is mainly due to the amount of insulation used. Not very high-tech. In fact, you may have seen pictures of old refrigerators with large, round turrets on top. This was the compressor, and from a purely engineering standpoint, this made a lot of sense, since this is where a significant amount of the heat was dissipated. Today, we hide the assembly behind and below the unit, forcing the heat to struggle to escape.

∞

As we've seen, electricity makes a motor turn by creating a temporary magnetic force in a coil of wire (e.g., in the rotor armature) that's attracted either to a fixed magnet or another electro-magnet (e.g., in the stator). We craftily keep the magnetic attraction moving ahead of the rotor's rotation, like a greyhound following (eternally) after the plastic rabbit at the dog races.

Although the rotational motor is by far the most common example of electrical energy doing work by moving things in the home, it is not the only one. Recall that the reason a motor rotates is that it is the only practical way to achieve continuous movement in one direction. If the desired direction is a circle, as with a fan, then the rotation is directly used. If the desired direction is a straight line, as with a garage door opener, then we use belts and pulleys to translate the rotational motion into linear (straight line) movement.

But not all movement in a home is continuous or, if linear (like the garage door opener), necessarily a long reach. Maybe we need movement that's only a half inch long. Maybe we need to move that half inch to tap a chime. As in a doorbell. Hark back to our original electromagnet, coils of wire around a metal core. Imagine that the metal core is an iron bar that can move freely back and forth within the coil. Now attach a spring to hold the bar just outside the coil. You can see that if we apply electricity, the resulting magnetic force created by the coil will pull the bar inside, stretching the spring. When we remove the electricity, the bar will bounce back. If we add a strike-plunger to both ends of the bar, and a chime plate in each side as well, you can see that applying electricity (pushing the doorbell button) will cause the plunger to strike one bar, and when releasing the electricity (lifting your finger from the doorbell button), the plunger will strike the other bar. Ding-dong.

When used in this way, the coil and plunger are called a solenoid.

Solenoids have another application if your home happens to include an outdoor sprinkling system. Electrically-activated solenoids open the water valves. This is not done with 120 volts, however. That would introduce an element of excitement every time you stepped outside when the ground was wet. No, instead, your 120 volt house voltage is transformed down to a more benign 24 volts (usually with a ubiquitous wall adapter). In the same fashion, a 12-volt transformer typically drives your doorbell. We'll revisit the whole subject of transformers in the next chapter.

There's yet a third type of solenoid in every home, at least, any home that has any sort of speaker, be it a stereo system, or a dinky little buzzer in your computer. Here, though, the movements are reversed—the metal bar is stationary, and the coil moves. Imagine that the inner bar is actually a magnet, and that the coil is suspended just at the end of the magnet bar. As we apply electricity to the coil, its magnetic force pulls against that of the fixed magnet, moving the coil inwards along the bar. The suspension of the coil has some spring to it, so that the more electricity we apply to the coil, the harder it pulls against the suspension, and the farther it moves in along the bar. If the electricity we're applying to the coil is changing in amplitude, then the coil will bounce back and forth, dancing in rhythm with the electricity. If that electricity is a signal produced by a microphone and amplifier, then the movement of the coil will mimic the vocal chords of the person holding the microphone. If, finally, we attach a light paper cone to the coil, the cone will push air back and forth in synchronization with the original vocal chords, reproducing the voice. That is a loud speaker.

In reality, our fixed magnet bar is actually both an inner bar inside the coil, and also a concentric ring around the outside of the coil, the two being of opposite magnetic polarities. The coil dances back and forth inside a narrow gap between these inner and outer magnets. There's not a lot of extra room in there. If you play music too loudly and exceed the specified limits of the speaker, you could deform the coil or the suspension and then the coil begins scraping against the magnet. That's the iconic sad scratch of a blown speaker. If you *really* exceed the limits, you could melt the wire in the coil, and your speaker goes completely dumb.

Note that a tweeter is just another type of loud speaker that's formed specifically to optimize the reproduction of higher frequencies.

In electronics, a loud speaker is a type of transducer. A transducer is any device that translates one form of energy into another. In common usage, though, the term is usually reserved for applications involving information rather than power, i.e., signals versus mechanical work. In this respect, a loud speaker is considered a transducer, since it translates sound (the signal) that has been encoded as electricity back into sound that consists of sound waves in the air. A clothes dryer, on the other hand, translates 120 volt, 60 cycle electricity into tumbled clothes, and the information content is zero.[4]

∞

This chapter was about doing work, moving things. But we ended up talking about how we might use electricity to do something other than work. This is a perfect segue into the next chapters, where we leave behind the mundane world of hefting and pushing things, to delve into an invisible world where we tease electricity into analyzing, measuring, monitoring, metering, controlling, and, yes, even thinking for us. Up until now, electricity has been our mule; soon it will become the mule driver.

But first, we take a small side trip to complete the picture of how electricity and magnetic energy are dance partners.

---

[4] The doorbell is actually a transducer, since, although a simplistic amount of information, the purpose is nevertheless a signal that someone wants your attention at the door.

5

Transforming Your Potential

We saw how coils of wire create mutually supportive magnetic fields that, when all added together, are strong enough to move a metal bar inside it (or move itself against a metal bar, as in a speaker). We learned that the electromagnetic field is a result of electrons moving inside the wire.[1] It turns out that it works both ways; a magnetic field changing around a wire causes electrons in the wire to all shuffle off in one direction, which we call current. The technical term for this is electromagnetic induction, and you might want to practice saying it, since it has a certain *je ne sais quoi.*

If I coil up a wire and swing a magnet past it, I get a quick burst of current. If I swing the magnet back and forth like a maraca player in a punk rock band, I'll get a current that swings back and forth in rhythm. This is, as we've seen, what we call AC current. Here comes the slick part. We know

---

[1] To be precise, moving electrical *charges* produce a magnetic field. They can be free negative electrons, or positive protons, and they don't have to be inside a wire. This is why the charged particles of the sun's solar wind (plasma) are funneled by the Earth's magnetic field to create aurora borealis far to the north (and aurora australis far to the south). It's also why a nuclear burst high in the atmosphere can cause an EMP (electromagnetic pulse) capable of knocking out electronics for hundreds of miles.

that an electromagnet works the same as a fixed magnet, and that an electromagnet is just some coils of wire with electricity passing through. So, I could, for example, hook a DC battery up to a coil of wire and use that as my maraca magnet. The key is that the magnetic field is changing around my first coil—let's call it the "target coil"—as I swing my hand back and forth. That's what's inducing the current. I cease my tiring hand-waving, and place the electromagnet coil—let's call this the "source coil"—right up against the target coil. Nothing happens in the target coil because, even though the DC battery is making my source coil into an electromagnet, it's not moving past the target coil. Are you ready? I replace the DC battery with an AC current—maybe household AC. My two coils are fixed in place, but the source coil's magnetic field is swinging back and forth on its own. This is the same as when I was moving the fixed magnet past the target coil. The AC current in my source coil is producing an associated AC current in my target coil.

This is pretty cool, but we're not quite done. We've seen how the strength of the magnetic field in an electromagnet is proportional to the number of turns in the coil. Again, this works both ways. The more turns I have in my target coil, the more it wants to push current.

We have to pause just a moment. Notice how I carefully worded that last sentence. I didn't say "the more turns, the more current," but rather, "the more turns, the more it wants to *push* current." We don't want to get side-tracked with the details of this, so let's just say that there's a type of resistance that develops when we try to push alternating current through a coil (this resistance is called "inductance").[2] We

---

[2] Since you're curious about more details, here goes.- When we pass electricity through a coil, it creates a magnetic field. But that magnetic field surrounds that same coil, and the magnetic field and current are mutually supportive. The magnetic field wants to keep whatever current is flowing the same. If the current into the primary is decreasing, the magnetic field tries to keep the current flowing. Conversely, if the input current is increasing, the magnetic field resists. The magnetic field always opposes change, in either direction. This is the AC resistance we call inductance, and it is proportional to the frequency of the alternating current. When the frequency is zero, i.e., DC, there is no inductive resistance.

know that current through a resistance is voltage, and we're now going to switch from talking about current in the coils, to voltage *across* the coils. I know this seems like a bait-and-switch, but you'll have to trust me that you don't want to know all the gritty details, at least not yet.

Back to the turns in the source and target coils (using voltage instead of current). The more turns in the target coil, the more voltage that is developed across it. If I double the turns in the target coil, I'll get twice as much voltage. Let's say that my source and target coils are essentially the same, except for the number of turns in each. Let's further imagine that the magnetic "coupling" between them is ideal, meaning that all the energy I put into the source coil magnetic field gets picked up by the target coil. Now, if I have twice as many turns in the target coil as the source coil, then I'll have twice as much voltage across the target coil as compared to the source coil.

This is called a transformer, because we can transform one voltage (it has to be AC) to another. We should switch to the proper terminology: what we've been calling the source coil is called the "primary" side of the transformer, and what we've been calling the target coil is the "secondary." If I have, say, four times as many turns in the secondary as the primary, then if I apply 10 volts AC to the primary, I'll see 40 volts AC (same frequency) across the secondary. We call this a 1:4 transformer.

It can go the other way. If I have one-fourth as many turns in the secondary, and apply 10 volts AC to the primary, then I'll see 2.5 volts AC in the secondary. This would be a 4:1 transformer. This is what's done in wall adapters. They're simply step-down transformers. A 10:1 "step down" transformer adapter would deliver 12 volts if plugged into a 120 volt AC outlet.[3] This could be used with the doorbell of the previous chapter.

---

[3] DC adapters add a few diodes to "rectify" the voltage and then optionally filter it. Some expensive DC wall adapters are moving to "switched mode" operation, which is more efficient, and can be made in much smaller packages—more on this in a later chapter.

Transformers can provide an additional feature when used with household 120 AC. Notice that there is no direct electrical connection between the primary and secondary coils. I can touch either side of the secondary output, and the only voltage I'll experience is what's directly across the two secondary outputs. I am making no connection to the potentially deadly 120 volts of household electricity, or even earth ground, for that matter.

Household appliance manufacturers have had a field day with these wall adapters. Any time an appliance uses the household 120 volts directly, it must go through a rigorous series of safety testing to be certified by UL (Underwriters Laboratories).[4] If they use a wall adapter, on the other hand (and the output is less than around 24 volts, which is virtually always the case), then only that adapter device has to be UL certified, and the appliance manufacturer is off the hook (at least as related to high voltage safety). The manufacturer of the wall adapter must go through the testing, but once approved, the adapter can then be used by any number of other devices.

When I started college, I knew that a transformer could step voltages up or down. This seemed really powerful, and I wondered why I had to have a complicated and expensive amplifier for my guitar. Why not just plug the guitar into the primary of, say, a 1:1,000 transformer, and connect the secondary side to big concert speakers? I guessed that there had to be a reason, and in due course (Fundamentals of Electricity 102), I learned why. With just a little thought, I should have realized that this would have violated the law of conservation of energy. Clearly the amount of energy contained in the feeble signal from my guitar is minuscule compared to that of pounding concert speakers—it's minuscule even compared to the sound tweeting from a small transistor radio. The energy delivered by the secondary of a

---

[4] UL certification is not mandatory, but if someone is hurt while using the device, and it's hasn't been UL certified, then the company might as well just hand over all its assets, because they're going to have a long, uphill battle in civil court.

transformer can never be more than the energy that's provided to the primary. Again, energy is everything.

We know that, from moment to moment, power is the measure of the instantaneous available energy, so we could expect that the power delivered by the secondary cannot be more than the power provided to the primary. The best we can hope for is that they are nearly the same. We also know that power is defined as voltage multiplied times current, $P=I*V$. Let's look at a 1:10 transformer as an example. If we apply 1 volt AC to the primary, we will get 10 volts AC out of the secondary. Let's further assume the transformer is a really good one—perfect, in fact. In that case the power provided to the primary equals the power delivered by the secondary. Doing the arithmetic, it's obvious that whatever current is delivered by the secondary to the load (e.g., the concert speakers), we will need ten times as much current driving the primary. If I connect my guitar up to the primary of a 1:1,000 transformer, and expect, say, one amp to be delivered to the concert speakers, then my guitar will have to pump out one thousand amps. Even Slash's guitar can't do that.

Conversely, if we use a step-down transformer, say the 10:1 doorbell transformer, then if the doorbell solenoid draws one amp, the household AC will only have to provide 1/10 of an amp.

This is pretty fundamental. High current at low voltage is the same as low current at high voltage. This is the main reason why power transmission is done at tens or hundreds of thousands of volts. A single hundred-thousand volt power line carrying one amp can provide ten amps to almost a hundred homes at 120 volts. Power transmission losses are mostly proportional to the current, so the higher the transmission voltage, the lower the current necessary, and the greater the overall efficiency.

Back to my guitar. We've accepted that I do indeed need an amplifier between my guitar and the concert speakers. That's what an amplifier does. It commandeers an external source of power to increase the power of an input signal. You plug an amplifier into an AC outlet, and it uses that power to boost the signal. A transformer has no external power source. We call that sort of electronic device "passive" for that reason. An "active" electronic component is one that requires power to be provided somehow.

Ah, but you may have noticed that tube amplifiers have monster transformers that drive the external speakers. This is because tubes are voltage amplifiers—the outputs are reproductions of the input signal with really big voltage swings. The speakers, on the other hand, "want" relatively lower voltages—typically around a volt or two.[5] The output transformers step down the high tube voltage to one appropriate for the speaker. Again, no power decrease involved—there's a lot more current available to drive the speaker than the tube needs to deliver.

So, in the end, I still needed that transformer, but it's just a bit-player, not the guy doing the heavy lifting (those are the tubes). Also, it's in the opposite direction—stepping the voltage way down, not up.

---

[5] Speakers are typically either 4 or 8 Ohms. Even when there's multiple speakers in a cabinet, they're usually configured in series and parallel arrangements to still present either 4 or 8 Ohms at their input. An 8 Ohm speaker driven with a fifty Watt signal draws 2.5 Amps ($I^2 = P/R$, or $I = \text{square-root}(P/R)$).

6

On and Off

Electricity Becomes Your Finger

The first and last things you do with an appliance is turn it on and off. This, of course, consists of applying the 120 volt house AC, or removing it. Traditionally—and frequently still—this consisted of a mechanical switch, typically a toggle switch, just like the ones that crowded the control panels of the Mercury, Gemini, and Apollo space capsules of the sixties. Your light switches are essentially toggles switches with fancier handles, whether the classic protruding levers you flip with your finger, or the smooth plate rockers you tap with your elbow. In the case of on/off switches, these toggles are termed SPST, which means, "Single-Pole, Single-Throw," which in turn simply means that the switch opens or closes just one electrical path (the single pole), and that the switch does nothing more than make or break contact between the two sides of the attached wires (the single throw). Again, whether a light switch, or an on/off toggle on your vacuum cleaner, the single electrical path of the SPST switch is the hot wire we described back in chapter 2.

Not all appliances require that you use your finger (or elbow) to turn them on and off. Many do it themselves—a coffeemaker with a programmable clock function that automatically brews the caffeine before you rise is a good example. In this case, your appliance must toggle the toggle switch on its own. If you're imagining a mechanical jointed finger inside the heating cabinet, then you made a wise investment in this book.

Prior to the invention of the transistor, turning equipment on and off automatically or remotely was done with a small version of the solenoid we just discovered called a relay. Instead of flipping a chime striker back and forth as in a doorbell, the electromagnetic force pulls a spring-loaded contact to make an electrical connection between two external terminals. You can see that this is essentially an SPST switch, except instead of being thrown by the force of a finger on a toggle lever, the contact is made by applying a current to the internal coil. Another difference is that, once the mechanical SPST toggle switch is thrown, it stays thrown, whereas the relay contact only persists as long as the coil current is applied.

You may be wondering how this achieves anything, since it still requires electricity in the first place and simply replaces one electrical connection with another—and you would be correct. The key is what form of electricity we're replacing. A typical relay that can handle up to 120 volts, and 10 amps, requires, for example, only 5 volts and 1/25 of an amp to activate the coil and switch it on. In terms of power (P=I*V), we can see that we can switch 1,200 watts with just 1/5 watt of control, a six thousand-to-one ratio.[1]

This relay could allow us to turn on your coffee maker with a few AAA batteries. More importantly, the relay attached to the potentially dangerous 120 volt house voltage

---

[1] One of the very first applications for relays was with the telegraph networks of the nineteenth century. After traveling some miles along wires, the exhausted Morse code signal arrives at the next station pitifully weak (or more technically, the resistance of the wire has reduced the voltage). If it has enough oomph to drive a relay coil, then it can instigate a much larger surge of current. Since Morse code consists of dots and dashes, which translate to simply bursts of current of two different lengths of time, the relay mimics, or repeats, the original series of dots and dashes. The weakened signal has been recharged. We can say that it has been amplified.

can be buried safely inside the coffee maker's base, while we're handling the harmless voltage from the three little batteries. In this case, the relay has achieved the safe remote control of the relatively large voltages and currents of the coffee maker.

I promised you not only remote control, but automatic turn-on. The relay is still the key component. The programmable timer in the coffee maker is part of a large (humongous) class of electrical applications we group together loosely as control circuitry. We'll get into this later, but for now we just note that control circuits, whether old-fashioned analog or currently ubiquitous digital, want to operate at relatively low voltages. Why? Why not? Low voltage (we're talking anywhere from a few volts up to twelve volts) is safe for clumsy humans to handle, but more importantly, the lower the voltage, the lower (by a squaring factor) the power consumed. Also, the higher the voltage we're flopping around, the more noise we're generating in the radio frequency (RF) range, which could perturb other devices, like remote controls and, well, radios.

So, just as it was convenient to use a lower voltage to remotely turn on and off the coffee maker, using a relay allows the low-voltage control circuit of the programmable timer to start the brew cycle while you sleep.

Although very small, the contact in a relay is still a mechanical switch, and it makes a slight tick sound when activated (or allowed to deactivate). You can tell when older coffee makers (or clock radios, or refrigerators, or any other control-activated appliance) turn on and off by the familiar tick.[2] But, no more. Relays have been replaced by power transistors, which, of course, are silent. Stealth control. We'll get into this soon, but before we leave the world of click-clacking relays, we'll review a few other points.

The relay I described above made the electrical

---

[2] The earliest computers—we're talking WWII—used relays (later, tubes). Although no more complex than a cheap modern kitchen timer, the incessant clicking of thousands of relays was nearly deafening.

connection between the external terminals when current was applied to the coil. In other words, the connection was *normally open*—this is, in fact, the formal classification of such a relay, and is indicated by the initials NO. The relay could easily be built so that the connection is broken when the coil is activated, in which case the relay would be normally closed—NC. Many relays have three external terminals. Let's call one of the terminals the common (which is indeed what it is usually called). One of the other two terminals would be the NO, and the last would be the NC. The common terminal connects to the NC terminal when the coil is not activated, and to the NO terminal when the coil *is* activated. Whatever voltage or signal is applied to the common terminal is switched to either the NC or NO, depending on whether the coil is activated. We compared the original NC relay to an SPST mechanical switch. This new, three-terminal, version is analogous to an SPDT switch—"Single Pole, Double Throw," where the double throw simply means that a connection is made in both positions. If we add a second coil, and a second set of NC and NO terminals, then we've emulated a DPDT switch—"Double Pole, Double Throw." Note that both "poles" are controlled by the same two coil inputs—we're switching two different circuits with the same control.

Finally, all the relays I've described so far are non-latching, meaning that when all control current is removed, they relax back their original connections. There exists a class of relays called "latching relays," which use two control terminals, and three "secondary" terminals labeled "common," "A," and "B." If current is applied to the "A" control terminal, the common terminal is connected to the "A" secondary. Even if the current is then removed, the common-to-A connection remains—it's latched. The only way to change it is to apply current to the "B" control terminal, so that the common is latched to the B secondary. If you apply current to both control terminals at the same time, you've made a mistake.

∞

In 1947, Shockley, Bardeen, and Brattain invented the first crude transistor at Bell Labs, for which they received a

Noble prize in Physics in 1956. They set the stage for a new world that has been evolving every since. Transistors launched the general category of semiconductor electronics, which have been replacing that of thermionics (tubes), and, of course, mechanical relays. Other than specialized industrial applications, seventy-odd years later, the replacement is essentially complete. Semiconductors control the world, from your TV remote, to jet airliners that can essentially fly themselves. Where Bell Labs' original transistor could be held in one hand, now you can balance a billion of them on the tip of your little finger. The advancement launched the whole science of miniaturization, an engineering concept that essentially hadn't existed before.

Let's take a look at this most basic semiconductor building block, the transistor.

We'll start off with an analogy. Imagine a river, over which the transport of bags of money needs to be carried. The bags arrive from the north, need to cross the river, and proceed on south. The problem is that at first there are no boats, and the money bags start piling up on the north side. A company manager arrives and finds that there are boats for rent, a dollar each. He gives the boat rental guy ten dollars, and ten boats head off, moving ten bags of gold from north to south. The problem is that the rental is only good for one trip. The manager has to give up another ten dollars to get another ten boats across. It still seems like a good deal, though—ten dollars to transport ten thousand dollars. And, as it turns out, ten boats at a time are able to keep up with the flow of arriving bags. In fact, it's more than enough, and there's some boats floating around unused.

We've analogized a three-terminal transistor. The north side of the river is called the "collector." Think of it as collecting the bags of money. The south side of the river is called the "emitter." Think of it as emitting the bags on their way. The manager providing the rental dollars is called the "base." Think of him as . . . the "basis" of all that's good with your job (*ahem*). As you've probably already guessed, the money being transported are the electrons comprising the electrical flow, but so are the dollars that the kind manager is feeding the rental guy. So: feeding a relatively small amount of money into the operation (via the base) results in a lot more being allowed to pass (from collector to emitter). In a way, it's rather like feeding a relatively small amount of current to a relay's coil, and allowing a lot more to pass between the two secondary terminals.

A transistor has the advantage over the relay in that it's not mechanical, so it can be made smaller (a whole lot smaller), and can be more reliable—no moving parts. There's another advantage. Whereas a relay requires a control current of sufficient magnitude to move a piece of metal, a transistor is only moving electrons. Where the example coil above required about 1/25 amp (that's 0.04 amps, or 40 milliamps), a transistor doing the job might need only a millionth of a amp (one micro-amp).[3] There's yet another huge advantage that we'll talk about later, but that I'll just mention it in passing here—the relay can't be switched just a little bit; it's either on or off.

On the other hand, the relay does have one advantage over a transistor. Current can flow in both directions through the relay (it's just metal contacts, after all), but current flows only from the collector to the emitter in a transistor. This is obviously a problem when switching the alternating current of household AC, but we wouldn't actually use a transistor in that case anyway—we'd use a close cousin called a TRIAC, which—when all is said and done—is just a collection of transistors configured to do just that (allow current to flow in both directions). In fact, the TRIAC was developed specifically for the purpose of switching AC current. It does have one drawback in that, like the relay, it is either completely on (conducting) or completely off.

A TRIAC could be used by your coffee maker's clock program in place of a relay to begin the brew, or to turn on any other appliance automatically. It is also used in a

---

[3] Actually, to switch 10 Amps at 120 Volts with just one micro-amp, you'd need three or four transistors in tandem, but since they're too small to see anyway, who's counting?

somewhat more subtle fashion, though, in your home as well. One or more of your lights may have a dimmer control, where turning the knob down dims the light. We might guess from chapter 3 that one way to do this would be to put a great big variable resistor in series with the light, but, of course, that would be silly. As you dimmed the light, the power being reduced is just picked up by the resistor. When you turned the light down low, the dimmer knob might be too hot to touch.

What we'd really like to do is to reduce the actual average current supplied to the light. Instead of having the power dissipated in a resistor, we'd like to just switch off the connection to the light for short bursts. See the connection? Since an incandescent (or CFL, or LED) light's brightness is proportional to the average current, we could, for example, use a very fast timing circuit to turn on a TRIAC for, say, varying amounts of time a hundred times a second. Any less often, and we'd start to see flicker. How less often can we actually go before we start to see flicker? It turns out, about sixty times a second.[4] That should sound familiar. Indeed, we can use the 60 cycle alternating current—the alternating positive and negative voltage—to perform the timing control for us.

First, we have to take a closer look at what this AC (alternating) house voltage looks like. It's sinusoidal. If you find yourself reflexively rebelling, as wretched days of tedious high school trigonometry class are dredged up from deep memories, well, relax. All this means is that as the household AC voltage goes up and down (positive, and then negative), it does so in a smooth, rounded fashion. You've seen pictures. In the old science fiction movies, the filmmakers often connected an oscilloscope to the line voltage to display a "futuristic" feel to the supposedly advanced science. Imagine that waveform as it curves up, over, and down, before doing the same in a negative direction. Now imagine a second waveform of the

---

[4] Not a coincidence. The early winner of power generation suppliers—Westinghouse— settled on 60 cycles specifically because he felt that this was the minimum before flicker in the early incandescent lights became noticeable. Edison was using 50 cycles at the time. Apparently his eyes weren't quite as sharp.

same shape, but delay it in time, meaning in your mental image, it is positioned somewhat to the right of the first one. Finally, imagine that your TRIAC doesn't turn on when the first waveform starts to rise but only after the second one starts its rise. You can see (in your imagination) that this cuts out a hunk at the beginning of the first waveform—the one supplying current to the light. That cut-out hunk represents energy not provided to the light, making it dimmer. You can also see that as we move the second waveform—the control one turning on the TRIAC—farther along to the right (later and later in time) that more and more of the first waveform gets chopped away, making the light dimmer and dimmer. The method we use to delay the second waveform involves principles that we'll get to later (namely, capacitors and phase delays), but be assured that it's easily accomplished using a simple variable resistor (which is what you're turning when you spin the dimmer knob). Note that the negative half of each cycle operates the same as the positive (just upside down).

∞

The theme of this chapter has been electrical connections that are either on or off, nothing in between. As important as this is, it represents, obviously, the most basic operation. Using just this mode—on or off—we could implement the Western Union telegraph system of the nineteenth century, but not Alexander Graham Bell's invention. A telephone requires the ability to communicate minutely varying fluctuations, for this is what a person's voice does. Bell's original patent described transmission of voice "by causing electrical undulations." If you've had the opportunity to see speech displayed on an oscilloscope, you might agree that "undulations" is a perfect adjective.

We'll next explore the infinitely varying world of "undulating" voltages. Welcome to the electrical realm of signals. We leave "electricity" behind, and move on to "electronics."

7

AC/DC

Telephones

I remember being slightly annoyed when I first heard about the band AC/DC. I was just starting college, and working diligently to absorb the intricacies of electronics. New terms were flying past my head daily, and I recall feeling a little smug that I was on the threshold of a world inaccessible (but actually just un-accessed) by most others. Along came this rock-n-roll band that blithely commandeered a fundamental electrical concept. Granted, it was *very* fundamental, but still, it should have been reserved for . . . well, for electrical engineering majors. At least the Young brothers of the band were born in Scotland, birthplace of James Watt, James Maxwell, and Montgomery Scott (you know, "Scotty" from, um, never mind).

As we know, AC stands for alternating current and DC for direct current. And, current and voltage are related by the load that they're driving, primarily resistance. In electronics we (maybe just me, but I'm writing the book) tend to think of operations in terms of the voltage where possible—the current follows from there. So, when we talk about AC versus DC electricity, we're really talking about alternating or direct voltage. A battery is a DC voltage source, and delivers current when a load (e.g., resistance) is applied. Your power utility applies 120 AC volts to your fuse box, and delivers current as you use it. This is just a convenient way of looking at it.

So far, the only AC electricity we've looked at has been the 60 cycle power utility feed. But this is not a usual type of AC electricity. It's too perfect. The power utility works very hard to keep that alternating voltage a nice, smooth, well-shaped sine wave. In nature, it would be hard to find such a perfect shape—a flute can produce a relatively pure sine wave, but it's the rare exception.[1]

What is common to all sounds in nature is that they *are* AC, i.e., they wiggle. Oscilloscopes weren't introduced until after Alexander Graham Bell died, so he never saw an accurate visual representation of human speech, his life's work.[2] Although you may have never seen speech displayed on an oscilloscope directly, you've probably seen it in books, TV, or movies. There's not much to make of it, really, just a squiggly meandering line. But what may appear like random wanderings of an ant, becomes music to your ears, or at least audible words. Your ears, coupled with your brain, obviously have something intricate going on.

A tone is a sinusoidal signal—a sine wave. The pitch of the tone is its frequency. Say the word "turtle" out loud, and it doesn't sound anything like a musical tone, but in fact the word can be thought of as consisting of a continuously varying suite of pure tones, all mixed together. Every millisecond— every microsecond—the compliment of tones changes.[3] When

---

[1] There's something of a contradiction in this. The universe actually loves things that are sinusoidal. It's by far its preferred mode. In 1822, Jean-Baptiste Fourier, a mathematician, French revolution activist, and compatriot of Napoleon Bonaparte, published a book on heat flow that included the proposition that (I'm simplifying a bit here) any signal, no matter how complex, is composed of a series of pure tones. It's just that we rarely find just one tone all by itself.

[2] Ingenious mechanical methods of capturing electrical signals were developed in his lifetime using photographic paper, but these only worked on waveforms that precisely repeated indefinitely (otherwise, the result was just a smear). The only waveform of that nature at the time were simple, pure tones, such as a flute makes. But scientists already knew that these were monotonic sinusoids, so the devices were singing to the choir, so to speak.

[3] If you say, "turtle," there's the part in the middle, when you're saying "ur," where the tonal variations are minimal. This is actually the difference between vowels and

viewed visually on an oscilloscope, what we're looking at is the sum, moment by moment, of all the sine waves making up the sound at that instant.

What I'm getting at here is the concept of frequency range. In that oscilloscope image of visually captured sound, the low frequencies are the slow ups and downs, and the high frequencies are the sharp jolts. The slope of the signal at any instant is an indication of its frequency components.

The human ear can detect frequencies from as low as 20 Hertz to as high as 20,000 Hertz (20 KHz), although the general workable range is from about 60 Hertz to perhaps 16 KHz. Unless you are an audiophile who has trained yourself to hear the minute differences, if you filtered out the frequencies below and above this range, you would struggle to hear the difference. In fact, MP3 music has no content at all above 16 KHz.

How about a telephone? The traditional landline telephone has a frequency range from 300 Hz to 3.4 KHz. Not exactly high fidelity. As it happens, however, the dominant frequency range of human speech—what the telephone was invented to convey—lies well within this range. You wouldn't mistake a person talking to you on the phone to one standing next to you, but you can usually recognize who it is. After all, the original intent of the telephone was to convey information—words—more conveniently than the Morse code of the telegraph system.

The telephone system specifically restricts—filters—your voice to this range for very practical purposes. At the low end, the frequency range below 300 Hz is used by your phone's ringer bell. Okay, your phone doesn't have a ringer bell, but the function is the same. When you make a call to your friend, the phone exchange at their end sends out a 5 Hz, 75 volt signal, which, in the original phones, banged the ringer bell, but in virtually all phones now is detected by the phone's circuitry, and the phone then does something to alert you—

---

consonants. Vowels essentially ride on a constant set of tones, and consonants tumble out in a frenzied mix of them.

beep, toot, or jangle. The 5 Hz, 75 volt values were carefully chosen to best bang the bell on the called phone, and are now legacy remnants of a bygone era.

By the way, you hear the ringer tone at your friend's end as you're waiting for them to answer, right? You know I wouldn't ask if it was actually so. No, the phone exchange at your end lets you listen to a little ding-a-ling that it's making up, just so you know it isn't ignoring you while you wait. That ringing you hear has nothing to do with what's happening at the far end.

The high end of the frequency range—above 3.4 KHz—is filtered out in order to conserve bandwidth. Prior to the digital revolution, long distance phone calls were frequency-multiplexed (referred to as frequency division multiplexing), meaning that many simultaneous phone calls were stacked atop each other up the frequency spectrum (via what we call frequency modulation), each with its own 3.4 KHz (actually, 4 KHz) band. Virtually all telephone transmission is done digitally now, but the value of bandwidth remains.

When you pick up the handset of your (landline) phone, you hear a dial tone, and you might have the idea that it's there all the time until you begin to dial. In fact, your local exchange detects that you've lifted the phone's handset and only then places the dial tone on your line, letting you know that your local service hasn't (yet) gone bankrupt. It knows that you've picked up the handset because it continuously drives your phone lines with 48 volts DC (actually negative 48 volts, for esoteric reasons of reducing corrosion), and when you lift the handset—going "off hook"—you complete the connection and the exchange detects the resulting current flow.

This 48 volts DC also powers the mouthpiece—what you talk into. I should say, it powers the mouthpiece if you still have an old, pre-electronic, phone (a cordless would *not* be one of those). Until the flexibility of electronics replaced it, all phones used a mouthpiece (i.e., a microphone) that consists of carbon granules. This was one of Edison's inventions that helped swell his fortune. His carbon particle microphone was significantly louder than the existing ones (one of which—the dynamic electromagnet type—we now use almost exclusively again). The only problem was that it needed to be powered. Thus the 48 volts.

Early (very early) telephones required an operator to make the connection for you—think Lily Tomlin on the sixties TV show, *Laugh In*. You yelled into the mouthpiece until she (it was always a "she") heard you. In order to get the operators back in the kitchen where they belonged,[4] telephone companies took advantage of the existing 48 volt powering, and invented automatic pulse dialing. If you're old enough, you may have used a dial phone; if not, you might find one in a museum. The actual dialing occurred when you let the dial go after you'd spun it. As it rotated back to the idle position, it would make and break a connection to that 48 volts as many times as the number you were dialing. In your local exchange, mechanical switch mechanisms jerked along in concert with the make-and-breaks. Later, digital circuitry counted the make-and-breaks electronically. DTMF (Dual-Tone, Multi-Frequency—AKA Touch-Tone) was introduced in the sixties, using tones instead of connection make-and-breaks. Today, when you "dial" a number on your smart phone, what was once such a major portion of a phone's function is now just a handful more bits in the digital flurry of information that flies back and forth between you and a cell tower (or home WiFi, or mobile hotspot, or maybe a hyperspace wormhole exchange next year).

There's one aspect of the older landline phones that many people don't appreciate. Because they're powered by the 48 volts from the local exchange, they continue to operate even when your power goes out—even when the power goes completely out across your whole county (the local exchange continues to operate using stacks of batteries). This is true for both the original dial phones, as well as the newer touch-tone versions, but possibly not if your base station hosts a cordless. Of course, most cell towers have battery backup now, so your

---

[4] A joke.

cell will probably continue to work in a power outage as well.

We've arrived at cell phones. These are essentially very sophisticated two-way radios that can connect into the existing telephone network (and, increasingly, into the internet). There was an important point there. When you make a call to a friend, both using cell phones, you can easily have the impression that you are talking cell-to-cell, but this is not the case. Every call you make, even if it's from one cell phone in your house to another in the same house, gets routed from the closest cell tower to a local telephone exchange (via what's called a backhaul). So don't badmouth your regional telephone companies too loudly.

The "cell" in cell phone refers to the way the service is geographically configured. Each call uses a slice of the limited available bandwidth associated with each cell tower, so there are only so many calls that can be made from any tower at the same time. The number of simultaneous calls is increased across a multi-tower region by re-using the available radio frequencies. This is done by carefully controlling the transmitted power (and often the direction) from the tower. A region is divided up into "cells," much like the honeycomb structure of a bee hive. If the transmission power of a tower is only just strong enough to reach the edge of its cell, then it won't interfere with a cell two cells away. That cell can use the same frequency. So the entire frequency band for cell phone usage is divided into sets, and towers are assigned frequencies such that no two adjacent towers share the same frequency. What happens in dense areas, like Manhattan? Make the cells smaller and smaller. This requires more and more towers (operating at lower and lower power), but more and more customers are paying for them.

Cell phones are by definition mobile (in Great Britain they don't call them cell phones—they're "mobiles"). Being mobile means that you could potentially travel out of a cell area. In fact, this happens continuously as you drive along a highway (of course, it's your passenger talking, not you). The cell service handles this by seamlessly handing you off from one cell to the next. You don't even know it's happening. This was one of the most challenging engineering aspects of cell phone deployment, and goes generally unappreciated.

Using cell configurations helps increase the number of calls available, but it's not enough. We've seen that energy is everything in the physical universe, but in the world of communications, it's bandwidth. If you can double your bandwidth, you can double your customers, and, correspondingly, your profits. This is done by squeezing every last bit of information out of every last Hertz in the band. The squeeze is, or rather the squeezes are multifaceted and very complex, involving mathematical compression of your voice, and intricate ways to multi-use frequencies. The company Qualcomm became staggeringly rich in a short time after the invention of one multi-frequency method called CDMA (Code Division Multiple Access).

None of this squishing comes for free. We all pay in the quality of the voice transmission. We've traded the predictable quality of our landlines for a barely comprehensible service. In exchange, we've gained the great convenience of talking while walking along the sidewalk and bumping into other people talking on their phones.

8

Music to your Ears

Stereo

In the 1940s every home had an AM radio, a source of news, but mostly entertainment. In the 1950s, "hi-fi" became king with the introduction of 331/3 records and FM radio. In the sixties, stereo was the definition of modern, and every respectable living room sported a console system with a turntable, AM/FM radio, and built-in speakers. And then the seventies partied into town with mirror balls and "The Hustle." When we weren't donning our platform shoes and fake silk shirts for a night at the disco, we were riding around talking on CB radios, or sometimes listening to music on our eight-tracks, but mostly struggling to un-jam the players. At home, we listened to our LPs or cassettes on our dated stereos, but spent a lot of time wishing we could afford quadraphonic systems.

Most of us had never actually heard quadraphonic music, but it was the next thing, and so, obviously better. Everything had only been getting better since WWII. In a way, the seventies was an awakening, when you found out that the X-ray specs advertised in the comics was just a hoax. After all, the seventies was Nixon, Watergate, the fall of Saigon, and the beginning of Iran hostages. And disco.

But at least we had quadraphonic music to look forward

to.

I never have heard it. Quadraphonics was emblematic of the tipping point, the bridge too far. Technology had overrun people's need, or even their enhanced pleasure. There are different theories about why quadraphonic music never took off: the four-channel media was too expensive; running wiring to four corners was significantly more difficult than just two; quadraphonic music had a single sweet spot, where stereo has a sweet line; even the fact that it was subconsciously eerie to have musicians apparently all around you. My guess is that it was a combination of them all (maybe not the last). The quadraphonic media would have come down in price eventually, but there just wasn't enough market to keep it floating. Quadraphonics over stereo just wasn't worth the culture shift. Listen to monophonic music, and then hit the stereo button, and the reaction is, "Wow!" as the music spreads out and takes on depth. Now add a four-channel amp and two more speakers behind you (and slip in the proper media) and the reaction is "Hmm. Oh, yeah. I think I hear the difference."[1]

Stereo has been the mainstay for fifty years, and is here to stay. We have two ears, not four. We want to sit in front of the band, not on stage with them.[2]

We all know that stereo uses two channels, and it's not a coincidence that it's the same as our number of ears. When a sound comes from directly in front of you, it reaches both your ears at the same time. But when it's located off to one side, it reaches the closest ear a bit sooner than the other. We're talking half a millisecond (the speed of sound times the six inches between your ears). We say that the sound contains a phase difference between your two ears, and our brains are extremely proficient at analyzing things of this sort. In our minds, we locate the position of sounds by their phase

---

[1] Or so I've been told.

[2] "Surround sound" is actually rejuvenated quadraphonics, but predominantly limited to public venues, such as movie theatres, and audiophiles who need to own something better than the rest of us, and who will probably now want a refund on this book.

differences.[3] Even though the sound is coming at us from all directions, in the end it's all stuffed through a small ear canal. If we place a microphone next to each ear and record what's coming at us, and then play it back into each ear, the sound entering the ear canal is the same, so we hear it the same. That's all there is to stereo—two channels capturing the differences in phase between the original sounds.[4]

The secret to stereo is all in the recording. Once accurately captured on two tracks or channels, the playback—from the media player, through the amp and speakers—is just "straight-through." As long as you've got two of everything, you're all set. That said, you should at least make sure that the speakers are matched. Of all the components in your stereo system, these add the most "color" (in other words, they're not completely linear). Differences in color will cause a bit of confusion in your brain's phase analysis, and muddy the stereo effect.

Speaking of coloring music, a recent retro trend has been to delve into vinyl LPs. As an engineer, I would find some irony in this if I didn't have fond memories of lovingly handling the plate-sized disks as a teenager. From the forties through to the eighties, reel-to-reel magnetic tape was *the* medium of quality recordings, and engineers fought a long, hard battle to maximize the reproductive accuracy of a mechanical needle bouncing around in spinning grooves. They never really got there, but the listening public got so used to the characteristics of vinyl records, it became "normal," to them. When CDs were introduced in the early eighties, the

---

[3] We generally think of phase as the time offset between two signals occurring within one wavelength—from 0 to 360 degrees. Above about 2 KHz, the wavelength of the sound is less than the distance between your ears, and the phase offset gets confused. We can locate the position of a complex high-frequency sound source because our brain's sound analyzers can track the time difference of the sound's complexities. But close your eyes and move a pure high-frequency tone around, and you'll have a difficult time locating it.

[4] I should note that simply varying the amplitude of a sound source in each channel can give a sense of position, but this method has little "live" depth. Early studio recordings in the sixties used it to exuberant effect, though, by having instruments bounce back and forth between your two ears as though the musician was being swung around in front of you on a long pole. Dig it, man.

really accurate reproduction was at first deemed cold and sterile. If you want irony, consider that just as we were appreciating the higher quality of CDs, high-tech miniaturization and the internet forced us to accept the compressed and lower-quality MP3 format. However, it didn't matter much, since everybody was using ear buds, which is like crumbling up the Mona Lisa and jamming her through your mail slot.

A word of caution, though, if you are thinking about re-visiting the wonders of fifties technology: vinyl disks are recorded using a frequency adjustment technique that reduces the amplitude of lower frequencies, while enhancing that of higher ones.[5] During playback, these recording adjustments must be reversed, otherwise the music will sound tinny, lacking low-end. This recovery process is called "equalization," and the precise parameters were formalized long ago by the Recording Industry Association of America (RIAA) so that everybody was doing the same thing. This is the familiar (unless you're under forty) RIAA "rolloff" curve reminder message that was printed on all album covers. This was why amplifiers of the era had special "phono" inputs—these inputs included the equalization circuit. Additionally, a turntable's cartridge has a relatively weak output, and the phono input included extra up-front amplification (called a pre-amp).

Many new turntables include the RIAA filter and pre-amp, and so are compatible with any amplifier. The problem arises if you use an older turntable with a newer amplifier that has (reasonably) let the phono inputs fall by the technological wayside. The solution is a separate unit that goes between the turntable and amplifier, and is called, variously, a turntable preamp, a phono preamp, or possibly even an RIAA preamp. They're not expensive—no more than

---

[5] The reason for this is that if not attenuated, loud low-frequency notes could swing so wide as to exceed the width of the groove, and by raising the level of the high frequencies during recording, when they are reduced during playback, so is the characteristic hiss of dust and scratches—all in the high-frequency range.

the cost of one new LP.

9

Let's Get Straight

One component that is common to all stereos—whether a clip-on MP3 player, or a hundred-watt concert-class behemoth—is an amplifier, as there are always speakers to drive. This is an amplifier's main, often only, purpose. In the case of an MP3 player, the amplifier is ant-sized, since it's only driving tiny earphone transducers.[1] For a home-theater stereo system, the amplifier may weigh twenty pounds, and make your neighbors grit their teeth as it batters your walls with a hundred decibels.

Let's take the case off and look inside.

If it's more than a few years old, you might find a fist-sized transformer near where the power cord enters. As we've seen in chapter 5, transformers transform one voltage to another. Assuming our autopsied amplifier is more than about five watts, this transformer will be stepping down from the 120 volt household AC to anywhere between fifteen volts (for a five watt amp) and forty volts (for the hundred watt

---

[1] The technical definition of a transducer is any device that converts one form of energy to another form of energy. In this sense, the definition is pretty broad, encompassing motors, antennas, thermocouples, fuel cells, and even a lamp. In general use, however, we limit the conversion to signals, i.e., varying physical forms that comprise information of some sort. Thus, a thermocouple communicates temperature, whereas the only information we get from a motor is that power is still being applied.

In the case of a speaker, of course, the conversion is from electrical to sound, and whether pounding rap music actually contains usable information may be a subjective call.

behemoth). This is the first step towards creating a DC power source for the rest of the amplifier's circuitry.

Virtually all circuits—certainly any that process signals, such as our amplifier—need a constant DC source of voltage for powering their operation. Remember that our goal is to take the weak wiggle (the input signal) from the media source (e.g., a CD player), and turn it into a big muscular wiggle, able to drive our speakers. If we tried to do this directly from the sinusoidal AC line, then, as the input power rose up and over its 60 cycle curved path, the amount of amplification would do the same. The output wiggle would start at nothing, and then rise to maximum, and then fall again—sixty times a second. And when the input AC went negative, it would probably burn out all the circuitry (I'm exaggerating for emphasis). The point being, the amplification circuitry needs a stable DC platform to work from.

Reducing the AC input voltage was the first step. The next step towards stable DC powering voltage is getting rid of the negative halves of the AC cycle. For this, we go back to the diode we first saw in chapter 3. You may remember that we noted that diodes only conduct current in one direction, sort of like the turnstile gates at the zoo entrance. We can use them here to let just the positive half of the 60 cycle waveform through. Using a diode, we've created a simple half-wave rectifier—instead of alternating current, we now have just positive.

Now we have only positive voltage, but it doesn't look anything like a constant DC level—it looks like a line of camels walking nose-to-tail. We need to smooth it out, average it. Imagine that you've made piles of sand all in a line along a beach, and now you take your cupped hand and run it sideways along the tops until all the sand is even. As you were smoothing the piles, the sand filled your cupped hand as you skimmed along the top of a pile, and then the sand fell out of your hand into the next trough. The electronic hand that we use is a capacitor. Just as your hand temporarily held the sand during a peak, the capacitor holds electrons, storing them during the peaks of the positive 60 cycle waveforms, and giving them up during the troughs.

Imagine that instead of using your hand to smooth the sand, you used a teaspoon. The spoon wouldn't hold nearly enough sand to fill the next trough, and you'd end up with a line of piles with just the tops knocked off. Instead, we'll try a tablespoon. Still not a smooth wall of sand, but better. Maybe a serving spoon. Even when you use your hand, some small amount of sand may not make it into the trough. The bigger your smoothing storage, the evener the final straight wall of sand will be.

It's the same with our capacitor. The larger the "cap," the smoother the final DC powering voltage. Remnant 60 cycle undulations in our powering voltage will be heard as the annoying, immediately recognizable 60 Hz hum, the bane of audio systems. This is what you hear when the ground connection on your input cable breaks, except that now you're hearing all the 60 cycle AC around you in the house as your input wire becomes an antenna.

It's difficult—nearly impossible—to get rid of 100% of the 60 cycle ups and downs, but by using a large enough cap, we can reduce it to the point where it's not noticeable. "Not noticeable" is, of course, a subjective assessment, and as you look at more expensive amplifiers—more dollars per watt—the power conversion capacitor will generally be correlatively bigger.

There's another reason why the input power capacitor needs to be large. Going back to the sand analogy, imagine that, as you're trying to smooth the sand peaks, your kids (or your little brothers) are taking sand away to build a castle. They normally take just a little at a time, but sporadically grab big fistfuls, so you get smart, and instead of just your hand, you start filling a bucket as you smooth. This way, when they come and grab extra handfuls, you use the sand you've stored in the bucket to keep the wall straight. As our amplifier feeds bursts of high current to the speakers (Roger Daltrey reaching a climactic moment), our capacitor needs to be a big bucket, able to give up as many electrons as needed. If it's not able to deliver, then that solid DC powering voltage sags. The result? *Distortion*—the curse of audio.

The process of using a capacitor to smooth out the positive halves of the 120 volt AC power input is a type of filter, and keeping the DC level constant is made more difficult when we have to filter from one positive cycle to the next. Our simple one-diode rectifier allows just the positive halves of the input sine wave to pass through, and we're clearly not taking advantage of all the available energy when we just ignore the negative halves. A clever use of multiple diodes, however, will allow us to use the negative half—in fact, we can flip the negative half on its head and make it appear to be another positive half, giving us a continuous stream of positive cycles. It takes four diodes to do this, and the configuration is called a diode bridge. The four diodes consist of two pairs. When the input AC goes positive, one pair lets the current go through the load (in this case, the rest of the amplifier circuitry) in one direction, and when the input goes negative, the other pair routes the two ends of the secondary side of the transformer secondary in a reverse direction, so the current flows through the load in the same direction as the positive half. This is called a full-wave rectifier.[2] Using both halves of the input AC allows us to use a smaller filter capacitor, which is a fairly expensive component, so a welcomed cost savings.

∞

Now that you've gained a nice understanding of the way DC power has been used in audio amplifiers for eighty years or so, we'll talk about the method that's replacing it.

The traditional approach we just covered uses three basic

---

[2] When the transformer has a center-tap on the secondary, we can get away with just two diodes. A center tap is just what it sounds like, we tap into the secondary coil halfway along its length. When the input (primary side) is at its positive cycle, one half of the secondary is positive with respect to the center tap, and the other is negative. During the negative half of the input, the polarities reverse, so there's always one side of the transformer that's positive with respect to the center tap. We put a diode on each end, connect the outputs together (we're "ORing the diodes), and feed that to the load. The other side of the load goes to the center tap. Note that each end of the secondary has only half the voltage level referenced to the center tap as between both ends, so we need twice as many total coil turns on the secondary to get the same rectified output voltage.

steps: 1) transform the voltage, 2) rectify the voltage, 3) filter the voltage to obtain a DC powering source. The rectification is just a few diodes—small and cheap. The size and expense is in the transformer and the filtering (that monster capacitor). The new method doesn't eliminate these two, but reduces them greatly in size (and consequently, in cost). The secret falls back to the idea that the final filtering is just an average of the rectified sine-wave input. One (major) reason the filter capacitor needs to be so big is that it has to hold the DC output voltage from one 60 cycle peak to the next (actually half way, since we're using both the positive and negative halves, but still). It must hold that DC output rock-solid for 8 milliseconds. That's not much in human terms, but for a capacitor, it's tough.

So the new method doesn't do that. First, it gets rid of that transformer at the input. We take the raw 120 volts right into a rather crude filter—a much smaller capacitor. The output of the small cap is only roughly DC—we haven't tried to make it really steady. It still has some ups and downs as the input goes up and down. But this is okay. Here comes the magic. We use a transistor to chop this up, and *then* we filter it to produce the rock-stable DC output. The secret is that we're chopping it at a much higher frequency than the 60 Hz input—more like 20 kilohertz (20,000 Hz), or even 100 kilohertz. It turns out that filtering is a lot easier at higher frequencies. Doubling the frequency allows us to halve the size of the capacitor. So going from 60 Hz to 20,000 Hz clearly affords a much smaller capacitor.[3]

You're wondering how we're able to get the final DC voltage level we need since we took out the input transformer, which was doing the voltage transforming. Your astute question illuminates the core principle of the new method. Imagine that the transistor is chopping the crude DC out of our input filter fifty-fifty—the transistor is on half the time,

---

[3] The ratio is 333, but we can't actually reduce the capacitor that much. The 60 cycle filtering is just part of the story—we still need to provide the sand that the kids are grabbing.

letting the current through, and turned off the other half. Clearly, the average of that is half the average of the crude DC (which is hovering around 120 volts). Now imagine that I let the transistor stay on longer. The average voltage out of the filter rises accordingly. If I turn on the transistor less than half the time, the average voltage out of the filter falls. Obviously, I can create any DC output I want from zero to 120 volts just by adjusting the on-time of the transistor. In other words, I control the output DC voltage via the amount of time the transistor is switched on. For this reason, this new method is called a switched mode power supply, or SMPS. Since I'm controlling the output DC voltage by varying the width of each pulse produced when the transistor is turned on, this type of control is called pulse-width modulation, or PWM.

Note that our transistor is either completely on, or completely off, just like back in chapter 6. This is the simplest way to use a transistor, but in this case, just the ticket.

I might have experimented to determine just the right amount of time to turn on the transistor each pulse to achieve the desired DC output, but virtually all SMPSs use a control circuit that monitors the output, and varies the transistor on-time accordingly. Because we're doing this, and also fundamentally because we're using a transistor at all, we say that this approach uses "active circuitry." The original approach (transformer, rectifier, filter) is considered passive—each component does what it does with no control input. Also, since we're monitoring the output, and feeding that information back to the earlier on/off transistor, we say that this circuit uses "feedback control."

It would have been nice to end the story there, but alas, rarely is life as simple as we'd like. Since we handle the inputs and outputs of our amplifier with our fingers, we need to make sure none of them have a direct connection to the AC input (for obvious safety reasons). Therefore we need to insert a transformer back between the switching transistor and the output filter. This could be just a 1:1 transform ratio, and is called an "isolation transformer" (you're isolating yourself from a jolting death). However—and it's a big however—this transformer is much, much smaller than the input transformer of the original method, for the same reason that our output filter capacitor is reduced—a much higher frequency.

A final note on SMPS. These were initially developed for applications where tightly controlled DC voltage levels are critical—your computers have been using them for some time. In an audio amplifier, the precise DC voltage is not important, just that it remains completely steady, as any perturbations manifest as noise or distortion in the audio output. The fact that a SMPS does have precise control is a free benefit.

∞

If your audio device works off of a battery, such as an MP3 or portable CD player, then you don't need any of this. You're starting from DC. The engineers who develop portable players have the advantage that they can create a design tailored for the targeted batteries (approximately three volts, whether it's two AAs, two AAAs, or a lithium cell). There are instances where the available battery DC voltage is not appropriate—your car being the most obvious. In this case, the device will have a SMPS. Instead of chopping up the roughly filtered 120 volt household AC, it simply chops up the automobile's 12 volts, usually to something smaller. When used this way, the SMPS is called a DC-to-DC converter. That said, a car radio or CD player loves the 12 volts to begin with.

It's even possible to create a DC powering voltage that's higher than the one you're starting from. The chopped up output of the switching transistor is essentially an AC signal with a fixed amplitude but a variable pulse width, and, being AC, can pass through a transformer. So, instead of using the 1:1 isolation transformer, you could use one that increases the voltage.

Finally, if we chop the car's 12 volts DC nice and even at 60 Hz, run it through a special filter that smoothes it into a sine wave, and step it up 1:10 with a transformer, we've turned the car's 12 volts DC into 120 volts, 60 cycle AC.[4] This is essentially what's done when you buy a inverter that allows you to operate your 120 volt house appliances in your car (or Volkswagen hippie van).

---

[4] It's not quite that simple. I wrote 1:10 for the sake of streamlined comprehension. I've been using the RMS (root mean square) values for the AC electricity so far. An RMS value is very useful, since it represents the amount of energy that would be associated with a DC level of the same voltage. If I rectify 120 Volts AC and filter it, the resulting DC level will be 60 Volts (assuming no losses). Handy. For a sine wave (like household AC), the formula is that the RMS value is 0.707 times the peak AC voltage, where the peak is the amplitude of one of the half-waves (e.g., the amplitude of the positive hump). Going backwards, the peak-to-peak voltage is $(2*V_{RMS})/0.707$. The actual peak-to-peak voltage of your household AC electricity is thus 340 Volts. No wonder it bites you.

10

Getting Bigger

Before putting the cover back on our amplifier, let's take a look at how it achieves its reason for being. We understand that our stereo amplifier takes the weak little signal from some audio source—CD player, tuner, phono input—and pumps it out to the speakers. Ideally, the signal we push to the speakers should look exactly like the weak input, only about a hundred thousand times stronger. This is not an exaggeration.[1] So far, we've seen how transistors can be very useful going completely on, and completely off—appliance on/off controls and switched mode power supplies. Their real talent, though, is when they're operating somewhere in between.

We saw in chapter 6 how a small amount of current applied to the base of a transistor allows a lot more current to flow from the collector to the emitter. For the on/off applications, we made sure that there was enough base

---

[1] We know that power is $P=VxI$. But $I=V/R$. Substituting for I in the first equation, $P=Vx(V/R)$, or $P=V^2/R$. Audio amplifier signal inputs typically have a 10K Ohm input impedance. Standard nominal line level voltage is 1 Volt (RMS). Therefore the power delivered to the amplifier by the CD player is $1^2/10K = 0.0001$ Watts. If the same amplifier is delivering 10 Watts of power to the speakers, then the power-out to power-in ratio is 10:0.0001, or 100,000:1. If the amplifier is 100 Watts, then the power gain is a million-to-one.

current to keep all the collector-to-emitter current flowing that was needed to be "on," i.e., providing all the current that whatever is connected wants. If we start with no base current (no current flowing from collector to emitter), and slowly increase the base current, we find that more and more collector-to-emitter current can also flow. However—and this is a key however—the current that can flow from the collector to emitter increases more quickly than the base current. For a given increase of base current, we see a larger (but proportional) increase of collector current. This is the definition of amplification.

The following point might seem obvious, but I think it's worth emphasizing. The transistor doesn't create any current. The current that we see coming into the collector is increasing only because the transistor is allowing more current to flow. It's as if when there's no base current (called cutoff), the collector and emitter are a really big (infinite) resistor, and as we increase the base current, the resistor gets smaller, until at some point (called saturation) the base current reaches a level where the collector-to-emitter resistance is zero (the transistor is completely turned on). The point being that the power is coming from somewhere else—typically the DC powering voltage we created in the previous chapter. The transistor is just manipulating external power as it mimics the changes in the base current.

The gain of a transistor is the ratio of a collector's current change compared to a change in base current, and is labeled the Beta parameter (also referred to as h_{FE}). The gain can range from 50 to 200 for small transistors, and perhaps 20 to 50 for high-powered ones. Note that three stages of amplification, each with a gain of 50, would produce 50x50x50 total gain, or a total of 125,000—more than the hundred thousand needed above. That's the good news. The bad news is that we would never design an amplifier with unrestrained gains in each stage. The reason is complex, but suffice it to say that by reducing the gain of each stage, we can greatly improve the overall performance.[2] Gain stages of ten are more practical, and five stages (10x10x10x10x10) would yield our desired hundred thousand. Still not bad.

As I'm sure you understand, we're looking at the basic operating principles of amplification, and there's many details to getting it all working properly—details that we're skipping over, since we're not out to design an amplifier. But there is one detail that I think is worth exploring, as it has a general application to many circuits. This is the concept of AC coupling and biasing. The signal from our audio source—a CD player, say—is fed to the amplifier by a cable that has three conductors: ground, and the signal for each stereo channel. Each stereo channel is, of course, an AC audio signal, and by AC, we mean that the voltage goes both positive and negative (with respect to the ground conductor). Here's the problem. A transistor only works in one direction. We can only feed current into the base terminal, and the resulting collector current only goes from the collector to the emitter. If we drive the base of the transistor directly with the audio signal, the transistor will turn on and amplify only for the positive parts of the audio input—only half of the signal.[3] This doesn't simply result in a lower gain, it sounds like crap.

What we need to do is raise the entire input signal—positive and negative parts—up in voltage, so that the midpoint of the signal (where the ground level was originally) now rides at the halfway point of the transistor input range. It will be all positive, but still wiggling up and down with the same shape as before. When we do this, the signal is technically no longer AC, but we tend to continue to think of it as so ("AC" versus a direct current with no information content).

So, how do we get it to ride up above ground? The first

---

[2] This gain reduction is done using negative feedback, which serves to provide a steadier and more predictable gain, reduces distortion, and improves frequency response.

[3] Further, and we're getting into nitty details here, the transistor doesn't do well when the base current falls towards zero. Specifically, the base-to-emitter voltage must be greater than about 0.7 Volts to turn on the transistor at all.

step is to decouple it from ground. We do this with a capacitor. We've already seen that a capacitor can store electricity like a battery, but it has another property as well.[4] A capacitor has a kind of resistance (reactive impedance, formally) that's dependent on the frequency that's trying to get through it. The higher the frequency, the lower the resistance. At an infinitely high frequency, it would look like a wire, and at zero Hertz (i.e., DC) it has an infinite resistance. It just doesn't let DC current through.[5] So, we simply put a capacitor in series with the input signal. The capacitor blocks the DC component of the signal, essentially allowing it to float away from ground. In fact, we say the signal is now floating.[6]

The next step is to lock this floating signal at the midpoint of the transistor input range. We use resistors to do this. Imagine that you're on a ladder, halfway to the top. You'll have as many rungs above you as below. Similarly, if I connect two resistors of the same value (analogous to the same number of ladder rungs) in series, and tie one end of the pair to my DC powering voltage (that we developed in the last chapter), and the other to ground, then the voltage in the middle is halfway between the DC powering voltage and ground. If the powering voltage is 12 volts, then the midpoint in my resistor ladder is at 6 volts. If the resistor on the top is twice as big as the bottom resistor (the top has twice as many analogous ladder rungs), then the voltage in the middle will be 1/3 of the DC rail—4 volts in our example.[7] By selecting

---

[4] It's not really a different property, but just how the common property, capacitance, is viewed—sort of like moving electric charges versus magnetism.

[5] And at any given frequency in between, the resistance—the reactive impedance—is determined by the capacitance value. At any given frequency, the higher the capacitance, the lower the reactive impedance. The actual impedance is given as $1/FC$, where F is the frequency in Hertz, and C is the capacitance.

[6] But we can't use any old capacitor that happens to be lying around. We need to make sure that our blocking capacitor is large enough to let the interesting low frequencies of the signal through. Like the power supply filter capacitor of the previous chapter, expensive amplifiers will have larger capacitors in these areas.

[7] This is called a voltage divider, and the general formula for calculating the voltage at the point between two resistors is $V*R_{bot}/(R_{top} + R_{bot})$, where R_{top} is the resistor on the top

the two resistor values judiciously, we can make the nominal center point of our input signal (what used to be the ground level) whatever we want, and what we want in this case is the middle of the transistor's input range.

To review: the capacitor provided decoupling (from ground), and the two resistors (a simple resistor network) provided biasing in order to position the signal properly at the transistor's base input.

We've been mixing currents and voltages a bit, and we should take a moment to untangle them. A transistor is a current mode device, meaning that, as we saw in chapter 6, we control the collector current by varying the (smaller) amount of base current that we drive into the transistor. Let's add another resistor, this time between our power rail and the transistor collector. Now as the collector current varies (according to our changing base current), the current in the collector resistor also varies proportionally, and, by Ohm's law, the voltage across it also varies accordingly. If, for example, the collector current goes down, the voltage across the collector resistor also goes down, but that means that the voltage across the transistor (from collector to emitter) has to go up by the same amount. This is because the sum of the two voltages (across the resistor and across the transistor) must equal the powering rail voltage. (Why? It's just one of those laws.) So, even though the transistor is a current mode device, we've effectively made it a voltage amplifier by pulling its current through a resistor.

Let's not stop there. Let's replace that collector resistor with our "load"—i.e., whatever we're supposed to be driving after this gain stage. Now the amplified base current is again "manifesting" as amplified current. We can go both ways.

The gain stage we just built using a collector resistor is a fairly simple structure, and it is commonly used in amplifiers, at least at the front end. Besides being simple, it introduces little distortion (when supplemented with a few tweaks, such

---

(connected to the voltage V—in our case, the DC power rail), and R_{bot} is the resistor on the bottom (in our case connected to ground).

as negative feedback). Its main problem is that it's not very efficient. This is fine at the front end, because we use the first few gain stages of amplification to get the voltage up, not the power level.[8] Eventually, however, once the voltage level is where we want it, we need to start amplifying power (in our case, adding current oomph to drive the relatively low resistance speakers), and this requires more complex circuit structures. Our simple one-transistor gain stage that we built is one type of a general class of amplifiers, and it's called "Class A." Circuit structures for amplifying power generally have dedicated transistors that work with just the positive and the negative portions of the signal.[9] The output power-gain stages are usually Class AB (there's a Class B, but it's not used in audio applications) or Class C. These are commonly referred to as push-pull gain stages (one half pushes the current into the load, while the other half pulls it, and then they swap).

Actually, Class C is usually not used for hi-fi audio amplifiers, because unless extraordinary mitigating precautions are taken, it exhibits a high amount of distortion (when the signal crosses ground, and one transistor is turning off while the other is turning on—this is called crossover distortion). This leaves the Class AB, which is the most common output stage.

I should mention that there's also a Class D type, which resembles a switched mode power supply more than a traditional audio amplifier, since the audio signal is chopped up, amplified, and reconstructed with an output filter.

∞

Before leaving the realm of amplifier circuits, we'll (very) briefly review some other general terms you may one day

---

[8] And even though we're increasing the voltage—substantially—the power remains pretty low because we're working with relatively high resistances. Since power is $P=V/R$, a high resistance means a low amount of power for a given voltage.

[9] The efficiency is achieved by not having to bias the signal up, where the transistor ends up dissipating a large portion of the operating power. By amplifying the positive and negative portions separately, there *is* (almost) no biasing—they're both starting from ground level.

bump   into:

bipolar transistor (or bipolar junction transistor—BJT)—this is the kind we've been working with, and the kind originally invented by Shockley and crew;

NPN: one flavor of bipolar transistor—the one we've been describing;

PNP: this other flavor of bipolar transistor operates in a similar way to the NPN, except that the currents are reversed—you control current flowing from the *emitter* to the collector by *pulling* current out of the base of the transistor (from the emitter). This is used in applications where we want to turn on the transistor with a "sink" current instead of a "source." It's also usually the other half of a push-pull gain stage;

FET (field-effect transistor)—works generally like a bipolar transistor, except that its input is voltage mode instead of current mode. As such, it has a very high input resistance, which is its star feature. Like a bipolar transistor, it has three terminals, but they're named differently: the "gate" is the controlling input (analogous to the bipolar base); the "source" is the amplified input (analogous to the bipolar collector); and the "drain" is the amplified output (analogous to the bipolar emitter). There are two main flavors of FETs—JFETs (older), and MOSFETs (newer). Digital chips (including computer chips) are almost exclusively comprised of MOSFETs;

Common emitter: the amplifier configuration we discussed in our example above;

Common collector (AKA emitter follower): in its most basic form, we simply take the resistor that was connected between the collector and power rail, and move it down between the emitter and ground. The effect is profound. Skipping over the gory details, this configuration serves as a current amplifier. The voltage out is essentially the same as the voltage in, but the current is greater. As such, it can drive a load with a low resistance. This configuration could be used as an output stage for small audio amplifiers, such as your

MP3 player driving the ear buds;

line level: the standard signal level for connecting home audio devices, nominally 1 volt. An interesting point—since all audio sources by design use this same level, there's really no difference between the various inputs of your amplifier. "Video," "CD," "Tape," "Tuner," are all identical, and you can mix and match at will. The main use is that the selected input is often displayed on the front panel, so it is a convenience for you if you keep them coordinated. The one exception, as we saw in chapter 8, is the "Phono" input, if your amp has one;

tube (vacuum tube): you know what a tube is, but we'll mention that they require high voltages to operate (to convince the free electrons produced by the thermionic emission to travel through space), hundreds of DC volts on the power rail for larger amplifiers. Other than the high voltages involved, a tube operates similar to an FET, where the FET's gate, source, and drain correspond to the tube's "grid," "anode" (or plate), and "cathode." They also have a heating element, powered by a dedicated 6 volts (necessary for the thermionic effect). Improvements to the operation required adding more sophisticated grid configurations, giving us the tetrode (two grids), and the pentode (three grids);

crystal receiver (crystal set): I couldn't resist slipping this in. With all the discussion of amplifying signals, it's perhaps worth remembering that in the early decades of radio—until tube amplification replaced them—these radio receivers were the exclusive means for households to listen to radio. Their beauty was that they required no power, other than what was received from the antenna. But best of all, they were free. In an effort to kick-start fledgling radio, the US government published simple instructions on how to make them in your own home using everyday materials. Their downside was that they only picked up strong, local stations, and you had to use sensitive earphones, not something you could cob together with odd parts in the garage. Despite these drawbacks, crystal radios introduced millions of Americans to the miracle of radio in the 1920s. I imagine that a family all sitting around a table, each listening to their own earphones must have resembled a modern family, except that back then they were all listening to the same song.

11

Tone Up

Many small audio devices no longer include tone controls. There's two reasons for this. The first is that digital recording and playback is precise, and essentially a perfect reproduction of the original recording. When we say that the result is "flat," we don't mean that some of the musical notes are a bit low in pitch, but that the reproduction doesn't include introduced emphasis or de-emphasis in the frequency response, i.e., no artificial tonal coloring. The second reason we usually don't find tone controls on small audio devices is that there's just no room. Okay, there's a third reason—the market competition is extremely stiff, and including more user controls would increase cost. Wait, there's a fourth reason—listening with ear buds, you probably couldn't hear much difference anyway.

On a home-stereo amplifier, however, we invariably do still find tone controls. This is partly because for a hundred dollars, or multiples thereof, we expect a certain complement of features, but also because with quality, full-range speakers and a variety of room acoustics, tone adjustment actually makes sense. Speakers mounted on the wall or placed on a shelf may need enhanced bass, while rooms with a lot of glass may be too "bright" (sound-wise), and require attenuation of higher audio frequencies.

We learned earlier that the pitch of a tone is its frequency, and this is true, but pitch and frequency are not exactly the same thing. Pitch always refers to a musical context, and in a simple sense, you could define it as the precise, specific frequencies of the notes of a musical scale. And, this would indeed be the case if all of our musical instruments produced pure tones, but then they'd all sound exactly like a flute (which, as we know, produces a pure sinusoid, when not blasted into fluttering rockasmic joy by Ian Anderson of Jethro Tull). When the sound is more complex, like, say, when bagpipes harmonize with a trumpet and harmonica, the underlying notes that you perceive may not be what electronic measuring instruments (e.g., a spectrum analyzer) would predict. This is because your brain is not (yet) an electronic measuring device. Our ability to appreciate nuances of music has evolved over millions of years, and an ability to recognized specifically spaced frequencies just didn't afford a survival advantage. Think of pitch as the intersection of the human perceptions of music and the measurable attribute of frequency.

The frequency of the A note above middle C on the piano is defined as 440 Hertz. The lowest note on a standard piano—low A—is 27.5 Hz. I stated back in chapter 7 that the workable range of human hearing starts at about 60 Hertz. This corresponds to around the ninth white key from the bottom on a piano. It sounds like a contradiction here,[1] but what's happening is that a lot of the quality of the notes played down there reside in their harmonics—complimentary notes that are higher in frequency, well within our "working" range. The lowest note available on a pipe organ comes in at 8 Hz, far below anybody's capability, and you hear this with your feet and chest, not your ears.

Putting musical notes aside, though, tone control clearly implies messing with frequencies. The tone in this case is not the specific frequencies of musical pitches, but the broader

---

[1] The pun was accidental.

sense of tonal color—the brightness of high frequencies, the "honk" of midrange, and the imperative power of bass. On early audio devices, the tone control was often offered via just one dial—turn it down, and the music emphasized bass, turn it up, and you accentuated the treble. At least, this is what they liked you to believe. What was happening in reality was that when you turned the knob down towards the bass setting, you weren't emphasizing anything, but rather, you were attenuating the treble part of the audio range. This made it sound as though there was more bass. Attenuating frequencies is easy—all you need is a capacitor.

Besides the ability to delay a signal, as we saw in the dimmer switch (chapter 6), and store electricity, as we saw in the DC power supply (chapter 9), we learned that a capacitor's effective resistance changes with frequency—the higher the frequency, the lower the resistance (AKA reactive impedance). In the last chapter, we learned how to use two series resistors to divide a voltage (when we biased our amplifier transistor). Let's do that again, but we'll replace the bottom resistor with a capacitor, and instead of the DC power supply voltage, we'll drive the top resistor with our audio signal. Now imagine that we choose a capacitor value such that its resistance at, say 1 KHz, is equal to the value of the top resistor. You can see that if we tap off the audio signal between the resistor and capacitor, we've reduced the audio signal by half when it's at 1 KHz. For higher frequencies, the capacitor's resistance is lower, and the audio signal is reduced further. Using the analogy from the previous chapter, there are more rungs now in the top part of the ladder than the bottom. At lower frequencies, the capacitor's resistance is higher, and the audio signal is reduced less. At really low frequencies, the resistance of the capacitor is so much greater than the top resistor, that we've effectively not reduced the audio signal at all.

You just constructed a low-pass filter. The resistor/capacitor combination lets the low frequencies through, but attenuates the higher ones. The higher the frequency (i.e., the more treble), the more attenuation.

We're not done. Imagine that we make the resistor in our series resistor/capacitor pair variable. As we lower the variable resistance for a given frequency, the resistance of the capacitor is greater—with respect to the resistor in our voltage divider—and the audio signal is reduced less. For that frequency, and all the higher frequencies, the attenuation is less. The effect is that we're filtering less of the treble frequencies. This is what happens when you turn our single-tone-knob up—you're letting more of the high frequencies through. Conversely, when you increase the variable resistance, only the lower frequencies produce enough resistance in the capacitor to avoid attenuating the signal. This is what happens when we turn the tone knob down.[2]

High-end early audio devices had two tone controls: one for treble, and another for bass—instead of just being able to attenuate the treble end, you could also attenuate the bass end. The treble tone control works essentially the same as the single-knob versions, and we've added the ability to also selectively cut out the bass end with the bass knob. Cutting the lower—bass—frequencies is done by simply reversing the positions of the series resistor and capacitor setup we've already created. Now attenuation occurs when the resistance of the capacitor is higher (as it is at lower frequencies). Again, there are more rungs in the top part of the ladder than the bottom.

These are simple, but perfectly workable, tone controls, and the fact that they are entirely passive (no active components, like tubes or transistors), made them attractive back in the day when a single transistor added some amount of cost to a product. By the seventies, though, transistors became really cheap, and tone control circuits became more elaborate. By adding amplification, transistors allowed selective frequencies to not only be attenuated, but also amplified. Also, since transistors could provide a boost, audio

---

[2] When we turn the knob up (lowering the top resistor), we say that we're increasing the filter cutoff frequency—the frequency where we begin to appreciably reduce the signal (where "appreciably" is down by 3dB, or 0.7 of the unattenuated signal).

signals could be sent through mazes of complex resistor/capacitor networks that provided accurate shaping of the frequency band—sharper cutoffs, and narrower, controlled bands. After exiting the arduous filter obstacle course, the weakened signal could be reconstituted with the transistor boost. In addition to bass and treble controls, we began so see midrange controls, and the tone knobs marked with both negative indications (on the left) and positive indications (on the right), since the controls really did reduce or increase the associated frequency band.

Transistors finally became small enough that many were crammed together into one package. The industry preferred the term "integrated," to "crammed," and thus we have the term IC, or "Integrated Circuit." An IC isn't just multiple transistors available for the circuit designer to use as she would, but entire blocks of transistor circuits shrunk down to one little chip of silicon. As such, an IC is a functional building block—a black box, if you will—for the circuit designer. Many applications had a broad enough market to warrant the high cost of developing specialized ICs—tuner blocks for transistor radios, for example. But a building block that was universal enough to be used in a variety of applications would have a broader market. Thus was born the integrated operational amplifier—op amp, for short. In a nutshell, an op amp has both positive and negative inputs, allowing it to accept differential signals.[3] It also has infinite gain (actually in the millions, but effectively the same). Let me pause a moment to make sure you got that. As it turns out, infinite gain isn't as impressive as it may sound. An op amp isn't intended to be used in this "open loop" mode.[4] Without diving into intricate detail, we use negative feedback

---

[3] A differential signal has two parts that are referenced to each other instead of to ground. The amplitude of the signal is expressed as the difference in voltage (or current) between the two conductors. This virtual signal can be both positive and negative, as the two conductors can be either positive or negative with respect to each other.

[4] Except in specialized situations where you simply want to detect that, for example, the positive input is higher than the negative input (the op amp output rails high), or the reverse (the op amp rails low). In this mode, we call it a comparitor.

to fix a precise gain. And it turns out that the lower we reduce the designed operational gain of an op amp (i.e., the more negative feedback), the better its characteristics. So, even though a single op amp could provide, say, a gain of ten thousand, a designer might prefer to use two op amps in tandem, each with a gain of a hundred, or even four in tandem, each with a gain of ten. The infinite gain just allows the circuit designer an easier job of working the feedback to achieve the desired gain.

Op amps, however, can be used in more complicated ways than just straight amplifiers. By using capacitors in the feedback loop, we can create filters with extremely precise parameters. Because we're adding amplification to the filter, we call these circuits active filters. These are the basis of the graphic equalizer, the audio buff's crown jewels in the seventies, and now often added to audio equipment almost as an afterthought. A graphic equalizer either amplifies or attenuates specific frequency bands, allowing a versatile tailoring of the overall response. The reality, though, is that for each person who "optimizes" the sound, the settings will usually be different. More often than not, a graphic equalizer just provides the consumer the idea that he's bought something sophisticated, and given him an opportunity to futz.

∞

Before leaving the world of frequency manipulation, we would be remiss if we didn't talk a bit about inductors.[5] An inductor is just a coil, although a metal core is usually added to enhance its operation. We first bumped into this in chapter 5, when we were introducing transformers. There, we recognized that the primary windings of a transformer present some amount of resistance to an applied AC signal. You can think of an inductor as the primary side of the transformer, with the secondary windings removed (or, equally, a solenoid, where the moving internal bar is fixed in

---

[5] Notice that when something is clever, the narrative voice is "I," and when there's the possibility of being remiss, it's "we."

place). Where a capacitor stores energy in the form of stashed electrons, a coil also stores energy, but in the form of the magnetic fields created by the current flowing through it. Imagine that we've applied a DC current to our coil, and the current has been flowing long enough for the resulting magnetic field to be set up. When we stop applying the current, the magnetic field begins to collapse. Here's where things get interesting. Remember when we learned that moving electrons in a wire create a magnetic field, and, conversely, a moving magnetic field induces[6] current to flow in a wire? Well, as the magnetic field collapses, it's effectively moving past the coiled wires, and therefore inducing current to flow. Further, the direction of the induced current is in the same direction as the original DC current. It's as though the coil is resisting stopping the flow of current, and this is exactly how we think of it.[7] A coil resists changes in current, whether the current is trying to decrease, or increase.

This resistance to changes in current manifests as a resistance to AC signals, which are, after all, nothing but attempts to create a changing current via an applied changing voltage. The AC signal is trying to pull the current one way, and the coil is resisting the change. The more quickly the current tries to change, the more resistance the coil presents. Thus, a coil presents more resistance to higher frequencies. For a given value of inductance (which is primarily determined by the number of turns of wire), the resistance to the AC signal is directly proportional to the frequency of the signal.

This is exactly the opposite of what a capacitor does. Inductors and capacitors are opposites, but complimentary to each other, and—along with plain old resistors—comprise the physical-world analogs of the dimensional, mathematical

---

[6] And thus I've snuck in the origin of "inductance."

[7] If you connect a coil across a small battery and quickly lift the wire from one end, you'll see a spark, because the coil is trying to keep the current flowing. This is essentially what a battery does, and we know that the battery presents a voltage in the attempt. Here, however, the attempt is, for a short time, much greater, and the resulting voltage can be quite high—thousands of volts—enough to create a spark across the gap.

abstracts describing the theories of electronics. Which we won't delve into. That's called an engineering degree. We will, however, make a note about the terminology. We've talked a good deal about plain old resistive resistance, and now we've introduced a different kind of resistance—frequency-dependant resistance, as manifested by capacitors and coils. These three "resistances"—plain old resistance, capacitance, and inductance—working together are collectively called impedance. The resistor is the resistive part, and the capacitance and/or inductance is called the reactance. When we introduce reactance to the circuit, we introduce time delays that appear as phase offsets. And with that, we close this chapter before you throw in the towel.

12

Heading Intruders off at the Pass

In 1990 I returned home from a pleasant evening with my then-fiancée (now my wife) to find to my surprise that I had left some drawers open in my bedroom. All of them, in fact. I quickly deduced that I would have remembered if I had done that, and, with the piercing logic of a modern-day Sherlock Holmes, concluded that my house had been burglarized. It's hard to anticipate the feeling that explodes in your head at this realization. "I can't believe it!" I shouted to the now-empty house. I repeated the proclamation over and over until I had to admit that, indeed, I did believe it.

The thief had taken my computer (in 1990, this was a severe blow), my electric guitar, and my watch. I went to bed, but lay in the dark, staring. *Was the microwave still there?* I wondered. I was sure I had seen it, but I couldn't resist getting up to check. Back to bed, staring. *What about the TV?* Up again, and then back to darkness and staring. This went on for a couple of hours and maybe a dozen trips, until I finally fell asleep. I was still a little groggy the next afternoon when the doorbell rang—the police had already come, lifted fingerprints (finding only mine), and left. It was a young man, maybe college age. Assuming he was selling something, I began the brush-off, but stopped. He seemed awfully nervous, and, glancing around, asked if he could come inside. Somehow I sensed that he was the one in possible danger, not me, and I let him in.

With difficulty, he explained that his friend had gotten drunk the previous evening and crawled through an unlocked window of my house and taken my things. He wanted to return them on behalf of his friend, but I had to promise to call off the police. I glanced out the window and saw his license plate perfectly visible, and thought that his jig was already up. He looked so distressed and forlorn, though, and I couldn't bear ruining his life for what was probably a one-time failing of judgment. For, it seemed obvious that the "friend," was really he. "Okay," I said, and stepped aside as he ran back out, opened his trunk, and quickly hauled my stolen stuff inside.

Seeming relieved to be rid of the damning evidence, he made for the door without delay. "Wait," I called. "What about my watch?"

He turned back, his face frozen in surprise and dismay. "There was a watch?" he asked, his voice cracking.

I realized that this young man had not been the thief after all. He was going way out on a limb to save his friend. "It's okay," I said, waving him out. Seconds later, he was in his car and on his way.

I may be the only homeowner who was ever robbed and had their stuff returned voluntarily the next day. But I am now more careful about keeping my downstairs windows latched.

And I now have an alarm system.

(I hope that I will be that kind of friend someday, if needed. The watch only cost $20, by the way.)

∞

Burglary prevention takes three forms: deterrence, detection/surveillance, and alarms. Deterrence, of course, is convincing the prowling burglar to move on. Classic approaches include leaving lights on, leaving lights on timers, leaving a car in the driveway, leaving a radio blaring, buying a dog (and leaving him at home), and maybe leaving a TV on. This last, in my opinion, is the most secure. There's nothing that makes it look like you're at home, ready to dial 911, than seeing that immediately recognizable flicker of light in the window. But leaving a television on, particularly a big flat screen job, is an energy waste, and it does shorten the life of this expensive and invaluable piece of equipment. Luckily, a little $25 device will simulate an operating television for a tiny fraction of the energy cost. It uses computer-generated patterns driving super-bright LEDs to perfectly emulate an operating TV—just plug it in, and point it at a window covered with blinds or curtains. But make sure it's the authentic FakeTV™ brand—the original, and the only one sanctioned by the inventor. The rest are cheap fake knockoffs with visibly degraded performance.[8]

Detection and surveillance fall partially in the deterrence camp, since when we do install outside motion-activated lights and surveillance cameras, we make sure they are visible. The message being, "You are being watched, bub." Surveillance cameras go beyond that, of course. We get to see who's at the front door, and whether it's really the pool boy heading around back of the house. Relatively cheap, compact, wireless camera systems have even brought the surveillance indoors. You can make sure the nanny isn't using your baby as a doorstop, and that your teenage daughter isn't using the clothes she was wearing ten minutes ago to block the crack under her door in order to keep the clouds of pot smoke inside, while she tries to convince her boyfriend to undress as well. In these cases, you've presumably hidden the cameras, otherwise you're just sort of testing their basic intelligence.

But, getting back to the technical side, there are three issues that you'd need to consider before investing in a video surveillance system, whether it's indoors or outdoors. The first is whether the cameras are wired or wireless. In either

---

[8] My editor tells me that I either have to take out the whole transparent FakeTV promotion, or explain here that, as the inventor of the FakeTV,™ I have a vested interest in convincing you to buy one. I chose the second option. (Go ahead and buy one.)

case, they'll still need to be plugged into an AC outlet. There are battery-powered units, but the inconvenience of changing out the batteries makes this option questionable at best, especially since the cameras are often mounted in high locations. The wireless cameras have come a long way in the last decade, and are now very reliable and easy to use. The convenience of not having to run new wiring through the house is tempting, indeed.

But, here's the catch, and it leads into the second issue—there's only so much bandwidth available with most of the wireless systems, and bandwidth equates directly with resolution. Wireless camera views are typically displayed on screens the size of your palm. The image is usually black-and-white (although some expensive systems are now color), and on these small screens, the resolution is not particularly objectionable. While you might recognize a friend, for anybody else the description could be something like, "He was wearing . . . a shirt—a long-sleeved shirt, and he didn't have a hat. I think." One clue that a wireless cameras has very limited resolution is if the manufacturers doesn't advertise the actual pixel count. More expensive systems may advertise, for example, 800 TVL,[9] but the tradeoff is in frame rate. What this means is that in order to get that much information across the wireless link, the camera will have to skip some number of subsequent image frames. This results in the characteristic jerky, stop-action type of display. You may be used to this if you were an early user of Skype, but you have to hope the intruder doesn't quickly hide his knife during the frame blackout.

If you want to see if your nanny is wearing ear buds when she's supposed to be listening for the baby, you will probably need the higher resolution and color of a wired system. There's almost no limit to the degree of resolution (other than

---

[9] TVL stands for "TV Lines," and literally indicates how many vertical lines can be displayed. These correspond roughly to the number of horizontal pixels, although only roughly, since TVL refers to analog systems, not digital. A TVL rating of 600 roughly corresponds to standard broadcast television resolution.

cost), and you can monitor multiple cameras with no loss of frames or resolution. The downside, of course, is the wires. Having them professionally installed could cost as much as the cost of the equipment. One alternative would be to duct-tape them along the baseboard and around door jams. Try to get that one past your wife (or husband).

The third issue is storage. Most systems now come with DVR capability, meaning that you can continuously record the camera feeds. Currently, using reduced resolution, a modest system can store up to a week's worth of continuous recording. After that, it simply starts recording over the oldest images. As time goes on, the available storage will undoubtedly increase, and with it the ability to store longer spans of higher resolution. Many systems also allow internet access for remote viewing and control, and, again, as time goes on, this will surely become a standard feature.

Home security alarm systems are more or less your last defense. If your deterrence methods didn't fool the guy that wants your stuff, and you don't happen to be sitting around monitoring the cameras at the time he decides to take it, then you're counting on alarm electronics to detect him and either scare him away, or get the police there to catch the SOB. I wouldn't recommend, for example, that you try to shoot him yourself, since over 98% of police alarm calls turn out to be false alarms. You know what they say: guns don't kill people—people accidentally kill people.

If you live in an apartment, and you're not on the ground floor, then your "system," might consist of just a door entry alarm. This could be as simple as a door contact switch that makes a connection when the door is opened, tied into a siren alarm. The siren alarm typically has a keypad that allows you to enable the alarm on your way out, and disable it on your way in. I don't need to remind you to mount the siren out of easy reach, otherwise he might just rip it off the wall.

If your home has both door and window access, then you obviously need entry sensors on each of these. Motion sensors are often part of the alarm suite, but keep in mind that they're more prone to false alarms, and, after all, wouldn't you like to shoo mister intruder away *before* he gets inside? But, whatever complement of sensors, your first decision is the same as with the surveillance cameras—wired, or wireless? If you're having your house built from scratch, you might consult with the contractor about having the wiring pre-installed. If he doesn't know what you're talking about, consider whether you've chosen the right contractor. Unlike video cameras that require either coaxial or high quality digital cabling, wiring your house for alarm sensors calls for just cheap twisted-pair wires routed around inside the walls. Make sure, though, that you get a wiring diagram from the contractor, even if he just draws it on his lunch bag.

If you're unlucky enough to buy an already-built house, then you're probably going to have to go with a wireless system. Fortunately, like the surveillance cameras, the reliability of these have improved greatly in the last decade. Worse than no system at all is one that false trips occasionally, shortening your life a few weeks with each event. In the nineties I made the mistake of piecing together a cheap system, and after a couple of months of repeated false trips, my wife finally asked that I remove it, stating that she'd rather the burglar take away what he could carry.

You will become a slave to batteries, though. You might as well buy them by the case. Manufacturers state that the batteries in their sensors last up to a year, but you should steel yourself to diligently do the rounds to change them at least twice a year—when you change your clocks for daylight savings time is a good time, since you'll have a reminder.

Security companies would love to install the system for you and then sell you on a monthly monitoring plan. You know how it works. When an alarm is tripped in your house, the system automatically makes a call to their monitoring center, and then they call your telephone to see if you've tripped it by accident. If they can't reach you, they call the police. Typically, you get two or three free false alarms, and then you have to pay a fee for each one after that. There's a certain appeal to these services, the sense that professionals are on the job twenty-four hours a day protecting you. In a way, it takes the responsibility out of your hands. But many of the new systems will automatically call your own telephone anyway. And now with cell phone technology, you're essentially on the job twenty-four seven anyway. In any case, the new systems give you the option of setting them for "silent alarm," where they just call you, or blasting an audible alarm.

If I were working for one of the security companies, I'd be training for another job.

Here's the thing. Our computers are getting smarter. One day soon, installing video surveillance cameras will be just the first step, setting up the eyes. Each camera will be smart enough to recognize that one or more people have entered the room, or your property. Further, the camera will also be smart enough to locate the face on each person, and will take a couple of zoomed-in shots of each, which it will then send off to some cloud intelligence (i.e., a server somewhere running high-powered software).[10] That cloud intelligence will match those faces to one of the millions it has on file—and there's very few people any more whose faces are not out on the web somewhere. The intelligence is very familiar with you and your habits, and it will start the search with people connected with you. If it fails there, it will expand to face matches within your geographical area. Once it has finished its analysis—upwards of a full second—it will contact you with the information.[11]

Each component of this scenario already exists. It's just a matter of time before it becomes affordable and ubiquitous. It

---

[10] This is a great example of the benefits of centralized cloud computing. The intelligence required to detect that a person had entered the field of view, and then locate the person's face, is modest, and easily accomplished by the commodity microprocessors already used in, say, your digital camera or phone. This is considered the pre-processing phase, and the result—a few megabytes of compressed image—is easily transported in a blink of the internet. Buried deep within a server farm, a powerful computer running powerful software then takes the image, chews on it for a few mega-flops, spits back the answer, and moves on. In one day, this muscular shared resource—just one of hundreds—handles perhaps a hundred thousand facial searches.

[11] I would say that the cloud intelligence would either call or text you with the information, but I fear that in five years, that sort of contact will be so old school. It might talk to you directly through an ear implant.

may sound like something you'd welcome in your home, but remember the other side of the coin—your face, your whereabouts, and your habits, too, will be (are being) tracked and noted. It's a whole new world, baby.

13

Nuclear Power for the Home

I live in California where silicon has been slowly coating the roofs of many houses. Sand is mostly silicon, but it's not the beach that's been blowing across the shingles. The perfectly geometrical forms appear on a house all in one day—in a matter of a few hours, in fact—and the homeowners couldn't be more thrilled.

I'm talking about solar panels, of course. Thanks to ever-rising electric utility prices, and government incentive programs, coupled with a burgeoning industry that rents the equipment, the glint (blinding glare, actually) of the morning sun is spread across any neighborhood, if seen from a high vantage point. As I write this, over 10,000 megawatts of solar power are feeding over a quarter million homes here. Over 5% of California's utility power is derived from solar—this is in addition to all those home systems.

The sun in California is clearly fueling more than just our tans.

Across the United States, the raw energy contained in sunlight averages about 100 watts on every square foot it falls upon. Averaged over twenty-four hours (night and day) and the four seasons (some rain, some sun), the long-term average is about a fifth of that, or more like 20 watts per square foot. Over the course of a year, that comes to about 175 kilowatt-hours. If the footprint of your house is, say 1,000 square feet, then over the course of a year, your roof absorbs (and reflects) 175 megawatt-hours of energy. If you're an average family, you use on the order of 8 megawatt-hours of electricity in a year. Based on this, it would seem clear that your do-nothing shingled roof is letting you down.

At least, these are the sort of statistics and conclusions the plethora of solar companies like to feed you as they're trying to sell their services. The reality is not nearly so overwhelmingly convincing, but still marginally suggestive.

First, solar panels are arrays of photovoltaic cells (AKA solar cells) that convert a portion of the solar energy that strikes them into electricity. Currently, the vast majority of solar panels sold world-wide are about 15% efficient.[1] That's an ideal, with maximum sunlight, shining on the panel at ninety degrees. Assuming this ideal, the original 100 watts per square foot of pure solar energy can be converted into 15 watts of electricity.

There is no ideal, of course, and the sun is actually nearly overhead for only about an hour during summer. Also, your roof is most likely pitched, and if you're fortunate to have one side facing south, the sun may strike it at an average angle of sixty degrees over the course of an eight-hour day. The solar panel's efficiency drops off as the incident angle decreases, typically down 6% at seventy-five degrees, and 10% at sixty degrees. So, during the course of the eight-hour day, at best, you might receive 108 watt-hours of electricity for every square foot. If your best pitch is facing east or west, then it's even less.

But that's assuming perfect atmospheric conditions. The normal amount of dust, moisture, and pollution reduces the

---

[1] Solar panels are currently being manufactured with efficiencies up to 25%, but the cost-per-Watt goes up very steeply as soon as you leave the mass-market 15% variety.

sun's intensity by about twenty-five percent.[2] That brings the daily watt-hours per square foot down to about 80.

You can't (yet) cover every square foot of your roof with solar panels, obviously. A typical installation might include twenty panels, each panel hosting 6.5 square feet of cells, for a total of 130 square feet of conversion. That brings us to a total of about 10.5 kilowatt-hours per day (average). Over the course of a year, that comes out to about 3.8 megawatt-hours of electricity.

That's just about half of an average family's usage, which isn't bad. There's a reason why you don't usually see homes with more than around twenty panels. First, many power utility companies have tiered systems of payment, and many of those that don't are moving in that direction. Tiered rates mean that you pay different rates per watt-hour, depending on how much electricity you use in any month. If you don't use more than the established baseline amount for a home, then you pay only the lowest rate—the "tier one" rate. Let's say that the baseline is up to 500 kWh (kilowatt-hours) per month, a typical figure. Further, let's say that one month you happen to use 600 kWh. For that month, you'll pay the minimum amount for the first 500 kWh (the baseline), and a higher rate—the "tier two" rate, often twice as much or more—for the extra 100 kWh. Each subsequent tier has an established threshold. The more electricity you use, the more your average per-kWh cost will be.

Now imagine you're contemplating installing some panels. A system isn't cheap, whether you buy it outright, or on a payment plan, or you lease it. A typical 20-panel system costs about as much as a decent car. You're calculating how long it will take to make back your investment in utility bill savings. Clearly, the biggest gain is keeping your monthly usage down, out of the higher tier rates. Let's say that your family uses a lot of electricity, and you're typically running 1,000 kWh a month. If you had enough panels to keep that

---

[2] That's based on my own measurements here in San Diego, where the air is dry and relatively clean. Where you live, you might expect more like 50% or less.

down to less than 500 kWh (about right for twenty panels), then you're effectively saving the cost of the tier two rate, a big savings. Anything more than that, and the returns are only at the tier one rate. Perhaps you calculate that you'll get payback on twenty panels in eight years, but not until twelve years for twice as many. Money talks.

There's a second reason that so many homes have twenty panels or less. Unless you're planning on going off the grid, power storage won't be included in your system. This involves banks of expensive batteries and associated control equipment. This could double your cost. But without a means to store excess power, what happens in the middle of the day when everybody's either at work or school . . . exactly when most of the electricity is being generated by your panels? Well, lucky for you, the federal government requires your utility company to have a means to allow you to put it back into the grid and sell it to them.[3] However, although the rate the utility company will pay you varies, it will never be more than their tier one rate. Once again, the payback equation barely balances.

Here's a few points regarding solar power for homes.

• A typical panel may be rated at 17 or 18 volts, but this is when it's unloaded (open circuit). The actual working voltage will be less, but generally above 12 volts;

• This is because the panel needs to potentially be able to charge 12-volt batteries;

• In which case, there will be a piece of equipment called a charge controller, which marries the varying output of the panels to the charge needs of the lead-acid battery;

• There will also be a piece of equipment called an inverter, which converts the low DC voltage of the panels (or batteries) to 120 volts AC for the house. This is similar to the auto inverter we talked about in chapter 9, only a lot bigger;

• This home inverter also has another job—being the

---

[3] The Public Utility Regulatory Policy Act (PURPA). –Different states interpret the law differently, and so are more or less friendly to selling back solar power to the utilities. California, as it happens, is one of the most friendly (consumer-wise).

referee between the grid power and that of the panels. The inverter must not only generate 120 volts AC from the panel DC voltage, but it must synchronize the resulting sine wave with that of the grid as well as match the grid's precise voltage;

• Also—and this is one that catches many homeowners by surprise—when the utility power goes down, your solar inverter is required by law to disconnect (to prevent your solar panels from back-energizing the power lines feeding your house, and thus posing a danger to linemen sent to fix the original power outage). When the grid goes down, so do you;

• Because of the way solar panels are wired (many cells in series), any shadow larger than one cell on any portion of one of the panels will likely de-activate the entire panel. Even a shadow across half of one little cell will reduce the output of the panel by half. It doesn't seem fair, I know;

• If you are about to sign a twenty-year contract with a solar company, look carefully at the fine print. Note what happens down the road if you decide to sell your house. Usually you are given the choice of either convincing the buyer to take over the contract, or you will be required to buy out the entire system. In the latter case, this could be covered by the sale proceeds, but would presumably increase your acceptable sale price. Either way, you'll have to hope that potential buyers are looking for a home with solar energy;

• Finally, the photon that was emitted from the surface of the sun and whacks into your solar panel, knocking out one electron, was originally generated by nuclear fusion at the sun's core. Thus the justification for this chapter's title.[4]

---

[4] And further, the gamma ray photon released during the fusion of two hydrogen nuclei took anywhere from forty thousand to a million years (there's disagreement about the exact number) to reach the sun's surface—through a process of absorption and re-emission that occurred trillions upon trillions of times while spanning the four hundred thousand-mile radius. It takes only eight minutes to jump 93 million miles to your panel.

14

The Digital Domain

Electricity Holds up its Fingers

The world is, by nature, analog, meaning it is changing continuously. Or, to put it another way, analog signals are representations of the real world. Everybody's probably familiar with the idea that digital means, to quote the Concise Oxford Dictionary: "relating to or using signals or information represented as digits using discrete values of a physical quantity . . . ." But our senses operate in an analog fashion. We don't hear things at discrete volume levels, nor only in specific discrete frequencies. Aromas vary as different smells mix and blend.

So why the fifty-year fascination and fuss over digital . . . everything? It's obviously not because a digital signal is a better representation of the real world. It's not. As we'll see in this chapter, though, digital signals have an overwhelming advantage in communication transmission, and early on we figured out how to do arithmetic and programmed control using binary numbers, the heart of what we call digital processing.

∞

Digital refers to using digits, and the word "digit" is derived from the Latin *digitus*—a finger or a toe. Our base 10 system of counting originated from the fact that this is how many fingers we have. Our distant, naturally clothed ancestor held up as many fingers as the number of rabbits he was able to kill. Then we invented the bow and arrow, and he ran out of fingers, so he then had to use multiples of hands-full-of-fingers (maybe using his toes to keep track). Thus, if he managed to inflict a rabbit massacre and came home with 26 rabbits, he could say that he had "two handfuls of fingers, and then an extra six fingers." Thus was born our ubiquitous base 10 system.[1]

As you are no doubt aware, our digital world uses binary numbers—just zero and one, instead of the zero-to-nine digits of base 10. We should note that "digital" does not automatically equate to binary. We could—theoretically—have digital computers that use a base 10 system instead of base 2 (binary). We use binary purely out of convenience. Implemented in electronics, base 10 becomes problematic when, for example, we need to have transistors operating at ten different discrete levels instead of two (on and off), which begins to degrade the robust nature afforded by binary operations—as we shall soon see.[2]

The first practical, widespread use of digital technology wasn't even computers, but rather telephony. After all, the transistor—the bones, tendons, and cells of computers—was invented at Bell Labs.[3] This digital technology wasn't something the general public was aware of, since it wasn't used in their phones, or even in any equipment within a mile of their homes. Making a telephone call from your house to your friend's house across town had never been much of a problem. It was getting your voice—and a thousand other

---

[1] Think of this as a History Channel dramatic re-enactment.

[2] If we were to start fresh, inventing the very first digital computer using the electronics technology available today, we might very well pick tertiary (0,1,2) or quaternary (0,1,2,3) numbering instead of binary, but above that, and the discrete handling starts to get clumsy.

[3] This was part of the original AT&T (AKA Ma Bell), once the largest telephone company—heck, the largest company, period—in the world, until its forced breakup by the US government in 1982. The original name was Bell Telephone Company—the "Bell" as in Alexander Graham.

simultaneous voices—intelligibly across the nation's landscape that bedeviled the AT&T engineers of the forties and fifties. The problem is that there is no such thing as an ideal transmission. As soon as your voice leaves your phone (translated into an electronic signal in the mouthpiece), it begins to degrade. Just a little until it reaches the local exchange, and just a little more until it's connected to your friend's home line, and just a little more until it reaches her phone, and just a little more until it's translated back into sound in her earpiece. But it all adds up. The "just a little mores" within your local area are perfectly acceptable. She hears you loud and clear (unless one of you is on a party line).

Getting from New York to California, though, was a different story. As we've seen, your voice would be frequency-multiplexed so that it could be carried along in its 4KHz frequency compartment along with hundreds of other calls. All this frequency modulation (and de-modulation at the far end) introduces some degree of distortion. That's not the worst part. The strength of the long-haul signal degrades with each mile—whether carried on a wire, or via microwaves—and the signal has to be received, amplified, and re-transmitted via repeaters. All of this processing, of course, causes some amount of distortion. Along the way, various sources of noise get mixed up with your voice, and the repeaters diligently amplify the whole party, uninvited noisy guests included. When you spoke to someone in California from New York in 1955, it *sounded* like you were talking to them from three thousand miles away.

The ultimate problem is that it's difficult (often impossible) to reconstruct an analog signal once it's been degraded. Let's take a simple example. Say we have an analog signal that communicates one of ten possibilities, perhaps a digit of a dialed telephone number. We'll let a value of zero be zero volts, a value of one will be one volt, etc., all the way up to ten volts. Let's say we're going to transmit a value of five—half the maximum. Resistance in the line droops the whole thing so that a one is now only 0.8 volts, a two is only 1.6 volts, and our communicated five value is no longer a robust five volts, but now a potentially masquerading four volts. Maybe we've expected some resistance droop, so we compensate, and are looking for something lower. However, another line may be twice as long, and the droop is twice as much, and yet another line, half as long, and the droop only half as much. It's getting difficult to predict what our "five" value will look like. Now add noise—have it bounce around from 3.5 volts to 4.5 volts, and we throw up our hands and declare the whole exercise futile.

Enter digital communication. Instead of a single line carrying the number as an encoded voltage, we now use four lines that represent the binary value for five. A binary five is 0101.[4] Each of the four lines communicates one of the binary digits. Since we're now only detecting a zero or a one, we can divide the ten volts into just two parts: anything above five volts we'll call a one, and anything below it is a zero. Now the line droop has to be bad enough that the signal level is cut completely in half before we begin to misinterpret the number.

Of course, we're sort of cheating by using four lines instead of one. We have another digital trick up our figurative sleeves, however. Suppose instead of four lines, I place the four binary digits on the same line, one after another. Let's imagine that we have some way to mark a beginning in time—a bell, or, better, another line that pulses just a moment to indicate the start. Then, let's say that each of the four binary digits occur on our single line one second apart. For each binary digit's one-second timeslot, the voltage is held at its value (high or low). At the receiving end, we see the start-time line pulse, and we start our clock timer. At one-half a second, we observe the line to see what the first binary digit is (at its halfway point). Then at one and a half seconds, we observe the line to see what the second digit is (at *its* halfway point). You see the scheme.

---

[4] A binary primer: binary numbers have increments just like base-10, where when we reach 9, we carry to the next digit for 10. Except in binary, the "9" is "1." So we count 0001 (1), 0010 (2), 0011 (3), 0100 (4), 0101 (5), etc..

We don't even have to stop there. Maybe there's another value we want to communicate as well, maybe the second number of the party we're calling. We can just place that four-digit binary value on the line right after the first one. At four-and-a-half seconds, we observe the line for the first binary digit of that number, etc.. We can keep laying in more numbers, one after another. The only limiting factor is the accuracy of our two clocks, the sending one that times when we place the binary digits on the line, and the receiving one that times when we sample (observe) the line. If the receiving one is running a little slow, for example, then it will be sampling a little later each digit, until eventually, we'll be looking at the wrong one.

In our exercise, we used time spans that we're used to—seconds. But in the world of electronics, instead of ticking away every second, we can use clocks that tick every thousandth of a second, or millionth, or, yes, billionth (the gigahertz clock speed touted in your laptop's ad). At a billionth of a second, you could cram (of course) a billion binary digits on the line in one second.

Clearly, we're talking hubba-hubba information capacity. This is called time multiplexing, and it's the core secret to digital communications.

Transmitting dialed telephone numbers is all fine and good, but we're talking about getting Uncle John's voice clear across the country. Just as we translated the ten levels of the original single-analog line into a four-digit binary number, we can do the same for John's voice. At regular intervals, we sample the amplitude of the signal, and translate the sampled level into a binary number. This is called an analog-to-digital conversion, or ADC. At the receiving end, we translate each arriving digital value into an analog level, recreating the original voice waveform in incremental steps. This is called—you guessed it—digital-to-analog conversion, or the pronounceable DAC.

That's the basic idea behind communicating a voice signal digitally—the complete process is a little more involved. First, instead of just ten levels, we typically sample the analog voice signal and convert the sampled level into one of 256 amplitude steps for a finer resolution of the voice quality. This requires an eight-digit digital number, rather than the four-digit we used for the original ten-level sample. Next, we make sure that we sample the voice signal fast enough that we don't hear the individual steps when we reconstruct it. "Fast enough" turns out to be a little more than twice the frequency of the highest frequency of the telephone voice band, or 8KHz (twice 4KHz). Finally, when we reconstruct the voice signal at the far end, we use filtering to smooth the steps. When done, the human ear can't detect that Uncle John's voice was chopped up, turned into a succession of numbers, and then pieced back together.

By digitizing Uncle John's voice, we've converted it into a form that can be carried around the world and back—to Saturn and back, if we want—with absolutely no degradation in quality. None. The only assumption is that the various intervening communication channels are at least good enough to discern between a one and a zero.

15

The Computer Revolution

Electricity Counts on its own Fingers

A computer computes. This is obvious, but our computers have grown so sophisticated that the process of performing actual mathematical calculations seems only incidental now. And, to a large part, it is. If, for example, a software programmer wants to include a logarithm calculation in his design, he can simply type "log,"[1] on a single line of code, and—when the program is later executed—the computer knocks off the calculation with a flick of its metaphorical finger. The same goes for square roots, trigonometric conversions, and exponentiation.

These finger flicks of a modern computer were originally their entire reason for being. During the eighteenth and nineteenth centuries—right up into the twentieth century—mathematicians, engineers, scientists, and navigators used published tables of these calculations, thick tomes of meticulously calculated values. These were hand-calculated by people who laboriously, hour after hour, day after day, performed the same series of arithmetic ciphering, not

---

[1] For the math savvy, "log" computes the natural log (to base e); to calculate the logarithm to base 10, the function call is "log10."

infrequently making mistakes. These people were called "computers," since they computed the values that inexorably filled out column after column of the tables.[2]

In the first half of the nineteenth century Charles Babbage designed machines intended to perform these calculations mechanically, but although construction for these were begun, for reasons of finance and personality, they were never completed, at least not in Babbage's lifetime. We know that they work, because museums have since made complete versions using his designs. Babbage then went on to develop designs for a machine that included all the basic components of what we consider a fully functional computer, including a means to program the operation, and, importantly, the ability to change its operation based on intermediate results. Even though these "analytical engines" were never built, he is credited as the father of the computer for the influence his designs had on succeeding generations.

Where Babbage's machines were intended for numerical calculations, another form of early computer reigned supreme for an important decade during WWII. These were mechanical analog computers, each designed for one specific application, and which used the continuous, interconnected movement of mechanical gears to model the problem to be solved. During WWII, the problems were gun control (big artillery and naval guns, not rifles), and aircraft bombing sighting. The Norden Bombsight computer was an example of the degree of complexity attained. Used to achieve bombing accuracy from high altitudes, it was capable of taking wind parameters into account to accurately predict the bomb's trajectory, and then automatically fly the plane during the final bomb run, and release the bomb bays. The development effort was massive—comparable to that of the first atom bombs—and so secret that bombardiers were ordered to

---

[2] If you're feeling sorry for these people, you might want to save it for the British coal miners of the time who developed debilitating rickets at a very early age because, working from before the sun came up, until after it went down, six and a half days a week, the sun almost never touched their skin.

destroy their units before donning their parachutes when the plane was going down.

These single-purpose mechanical machines were just beginning to evolve into much more powerful and flexible (i.e., programmable) electronic versions in the late forties and early fifties, when the advent of the digital computer sidelined further development. The Norden bombsight, though, continued its service through the Korean War and into the beginning of the Vietnam War.

Besides their inability to work with numerical-based computations, the analog computers lacked the capacity to change the course of their calculation based on their own intermediate results. The Norden Bombsight tracked continually changing parameters—wind speed, flight motion—but these were external inputs, not branching decisions based on its own calculation results. The power of a modern computer is derived from its ability to make its own decisions. It follows operational steps according to a program sequence read from a storage media that's flexible enough to be self-modified. A true computer re-programs itself, as was first described by Alan Turing, whose genius helped break the WWII German Enigma code, and thus, by many accounts, shortened the war by a few years. If Charles Babbage is the father of the computer, then Alan Turing could be considered the father of computer science.

∞

Prior to the digital revolution, any device that was programmed in the home was done with a clock timer, either mechanical, as in ovens and dryers, or electronic, as in clock radios. The only processing involved was counting seconds. That operation is easy to understand, and somehow we've managed to use the simple ones and zeros of binary numbers to enable you to dial your home and change the temperature in your house before you leave work. Just as your highly functional body is made up of trillions of individual cells that, on their own, can do little more than sit in one spot waiting patiently for some chemical handshake, the secret of the powerful capabilities of computers lies in the quantity of simple one/zero logic elements.[3] Binary operations are grouped together to perform simple basic functions, and these functions are grouped to perform more complex functions, and so on in a hierarchical fashion. A majority of the complexity lies in the stored program (the software code) that's directing the nanosecond-to-nanosecond operation, and this controlling complexity is possible only because of the transistor miniaturization that's been steadily progressing for the last half century.

Let's see if we can understand how we're able to dial our homes and change the temperature before leaving work. It does all begin with that one/zero binary operator.

The earliest numerical computational computer—a machine that could perform numerical calculations based on a stored program—was built using relays, two thousand of them.[4] This was quite a herd of click-clacking electromechanical components, but there's probably more binary elements than that in the display of your microwave oven. The relay was a logical choice at the time, since it's either on or off—a binary one, or a zero.

Tubes quickly replaced relays, which are electro*mechanical*, and the first all-electronic computer was Colossus, built towards the end of WWII by the British to decode ever-more complex German encryption (this was after Turing's enigma-breaking machine).[5] Colossus was a special-purpose machine—the first general-purpose computer was

---

[3] In fact, your body contains something like 30 trillion cells, give or take a ten-to-the-twelfth, and a 30 terabyte hard drive contains the same number of transistors. Of course, a single cell is maybe a thousand times more complex than a transistor.

[4] This was the Z3 computer, designed by German engineer Konrad Zuse just as WWII began, and used in Nazi aircraft design. In much of Europe (and obviously Germany) he is considered the inventor of the modern computer. The only working version was destroyed (unknowingly) by Allied bombing in 1943. Hitler refused to fund a replacement, saying that it was not important for the war effort.

[5] The five Colossus units, along with all the design material, were destroyed after the war, and the entire project kept secret until the 1970s. Sort of makes you wonder what else our governments are up to in secret.

ENIAC,[6] also built towards the end of WWII. Both Colossus and ENIAC were programmed using plugs and wires. The computer's instructions were literally wired connections, so re-programming was a project unto itself. The first general-purpose stored-program computers evolved through a few versions just after the war in Britain, culminating in the Ferranti Mark 1 in 1951. This was the last fundamental step towards a fully modern computer. The early post-war computers ran on instruction sets, and the programs consisted of sequences of instructions laid on paper tape. There was no "re-programming" the machine, just the next paper-tape program to be run.

The modern computer had arrived. It was big—the classic "it took up a whole room"—and, other than universities and government agencies, too expensive for commercial use. Also, the practical limit of tubes had been reached. The average life of a tube, divided by ten thousand tubes, meant that the operators crossed their fingers every time they kicked off a new ten-minute run, hoping the program would finish before the next tube burned out. Computers just couldn't get more powerful.

Just as the tube-wall was hit, however, Shockley and his crew's invention opened the gates for the future, and the race for smaller began.

∞

Sixty years later, smaller means billions of microscopic transistors switching on and off billions of times a second, but I still haven't explained how a binary on/off operator—whether a relay, tube, or transistor—can be made to execute program instructions, and thus change your home's temperature via your phone. Thank you for being patient.

Every modern computer—be it a smart phone, laptop, or

---

[6] ENIAC contained 18,000 tubes and 1,500 relays, weighed 30 tons, and consumed 200 kilowatts—enough power for a few hundred homes. It executed five thousand instructions per second, compared to your laptop, which performs nearly a million times that.

IBM mainframe—contains at its core a microprocessor.[7] After decades of ongoing miniaturization and concomitant increased integration, a microprocessor now has truly become a computer-on-a-chip. It still needs to communicate with the physical world—accept input from the user via a keyboard or touch screen, display outputs on a screen, interact on the internet—so it can't just float all by itself in space, but it contains, etched in its silicon wafer and protected inside its ceramic package, all the components of what we define as a computer.

At the core of its core (or each of its cores), every microprocessor executes program instructions one at a time. This is absolute. Whether a one-dollar PIC, or a thousand-dollar next generation Intel Xeon, each and every core executes just one instruction after another.[8] Putting aside the immense complexity represented by modern microprocessors for the moment, we'll explore a simplistic computer's program instruction execution for demonstration.

Every computer has an instruction set, and sorry, but we need to pause again. We need to understand what we're referring to. Computers have different levels of "instructions." At the lowest level, down inside the guts of the microprocessor where we are now, is what we refer to as "machine instructions." These are the instructions that the logic—the transistors—of the microprocessor work with one at a time. These instructions are fixed, determined by the design of the microprocessor. The microprocessor recognizes these machine instructions as individual binary words. We humans represent these binary words with character-based names

---

[7] It's fuzzy. Many microprocessors now sport more than one "core," and each core is indeed an individual processor engine executing its own stream of program instructions. Each core is not its own complete computer, however, since they all share external memory storage and I/O (Input/Output). If all four cores of a quad system can manage to not bump into each other, then such a computer could operate four times as fast as its single-core poor cousin, but this is often not the case, and normally one or more cores spend some amount of time waiting to get at a shared resource.

[8] Again, fuzzy. With pipeline instruction queuing and pre-fetch support, these architectures are getting closer to parallel processing within each core.

(called mnemonics). STA, for example, might mean "store a data value at this defined location." These character acronyms are called the assembly language, and the software program that helps manage them and turn them into binary words is called the assembler. Programmers rarely work at this level anymore, however. Instead, they use high-level languages, such as C++, BASIC, or Python, and the software program that turns these programs into assembly language (ready for the assembler to turn into binary machine instructions) is called a compiler. The compiler works within an overall computer-specific system management software called the operating system. It's important to understand that high level languages are the same, no matter what computer you're working with. The compiler's job is to translate it into the computer-specific language of that machine (i.e., the "machine" language).

Okay, back to our simplistic program instructions. Each instruction consists of at least one part, or field, called the opcode (operation code), which tells the microprocessor the basic operation it will be performing in this step. Here we have to pause yet again for a moment. Microprocessors work with fixed word sizes. Smaller microprocessors may use eight binary bits, laptop versions might use 64 bits. For our simplistic demonstration, we're going to use just three bits. Notice that three binary bits allows us up to eight combinations (000 to 111, or 0 to 7). The wider our microprocessor's word size, the more opcodes, i.e., different instructions, it can handle. Here's our set (that I made up):

opcode	operation
000 = LOAD	Load the register with a value.
001 = MOVE	Move a value from memory to the register.
010 = ADD	Add the value in memory to the register.
011 = SUB	Subtract the register value from that in memory.
100 = STORE	Store the value in the register to memory.
101 = JMP	Jump to a new instruction location.
110 = BRANCH	Jump to a new location only if a condition is met.
111 = JMPR	Jump to the location whose address is in the register.

Some explanations. Microprocessors can access (read from, or write to) binary values—the "data"—in different places. An obvious one is its memory, which holds multiple (many) data values, each one accessed according to its address, or its position, just like your house's address locates you on your street. Memory is further divided as local—inside the microprocessor—or remote—outside the processor—but our instructions don't care where the memory is physically located.[9] The other place that the microprocessor can access the data is local "registers." Think of these as tiny memories with just one location. They have no address associated with them, since there's only one data value. Registers are where the microprocessor actually works on the data values. Think of memory as the storage room, and the registers are your desk. In our tiny computer, we have just one register. Registers normally have labels, but we don't need one since we have just the one.

A complete microprocessor instruction generally consists of more than just an opcode. This is obvious, since, for

---

[9] The "local" memory can be further divided into local general memory—accessible to all the microprocessor cores—and "cache," which is specific to each core, and is located (functionally) really close to it, so the core can get in and out really fast. The cache itself is often even divided into levels—degrees of how really, really close to the instruction execution it resides (functionally).

example, "move a value from memory to the register" doesn't tell me which memory location to use. In this case, we need to follow the opcode MOVE with the location in memory. This is called the "operand." Let's say I want to move the value in location 26 of memory to the register.[10] The complete instruction would be:

MOVE 26

Similarly, a memory location operand is needed with the ADD, SUB, and STORE opcodes. An operand is also needed with the JMP opcode—where to jump *to*. Two operands are needed with the BRANCH opcode—where to jump *to*, and under what conditions to do the jump (e.g., the register value is zero or negative). Note that the JMPR opcode does not require an operand.

Looking at the LOAD opcode, on the other hand, we need to specify an actual data value, and this is called, unsurprisingly, the "data."

Finally, the opcode instructions (and operands, and data) are stored in memory somewhere, from where the microprocessor reads them, but we don't really care where the assembler program puts them. In fact, we use easy-to-understand labels in our program to tag different places (used for jumping around), and the assembler, smart as it is, converts these to the proper memory addresses.

Let's code up a snippet of program that checks to see if a particular value in memory location 95 is smaller than 8, and if it is, we'll add 7 to it, otherwise, we just move on. Here it is:

---

[10] We're using the word "move" here to be consistent with most assembly languages, but what we're really doing is copying the value from memory to the register. We're not changing what's in memory at this point.

label	opcode	operand1	operand2	data
start	LOAD			8
	SUB	95		
	BRANCH	positive	continue	
	LOAD			7
	ADD	95		
	MOVE	95		
continue	. . .			

Let's analyze it. At the location in memory labeled "start," the microprocessor executes the opcode "LOAD," which loads the value 8 into the register. Then:

SUB subtracts the value 8 in the register from whatever value is in memory location 95;

BRANCH jumps to the new program location "continue" if the subtraction resulted in a positive value, i.e., that the value in memory location 95 was 8 or larger;

(Note at this point that we branched away if the condition we're looking for is *not* true, i.e., that memory location 95 is larger than 8. Since we didn't jump, we know that location 95 must be smaller than 8.).[11]

LOAD loads the value "7" into the register;

ADD adds the register (7) to the value in memory location 95—the result is placed in the register;

MOVE copies the addition result to memory location 95, overwriting what was there.

This little code snippet in and of itself isn't very useful, but it does demonstrate how a computer performs minute, detailed steps, that, when strung together in their thousands, perform useful tasks. Perhaps memory location 95 contains a value that indicates the font size of one character on your screen, and this snippet of code is part of a routine that's

---

[11] This sort of inverted condition checking is common in assembly coding. Otherwise, we end up with more jumping around. If we had checked instead for the "true" condition, then we would have jumped to some place to do the subtraction, and then would have had to jump back—two jumps. This way, we just do one jump around the conditional processing.

changing the font size of your whole screen. This snippet might be part of a loop (maybe containing another few hundred instructions) that's run through for each character on the screen. Or it might be part of a trajectory program that's preparing an ICBM, directed at Canada, to launch. It's virtually impossible to tell what an assembly program is doing when randomly looking through the code. But I think it's a pretty safe bet you can eliminate the latter possibility.

I should note that, for the sake of simplicity, we've been using regular decimal values for the operands and data. In an actual assembly program, depending on the sophistication of the assembler software, these might need to be converted to forms more appropriate for the computer, e.g., hexadecimal.[12] In any case, the final assembled code—what's actually run in the microprocessor—is all binary.

This wasn't meant to be any kind of actual primer on assembly coding, but rather just to give you a general idea how a computer works, deep in its core. If you're interested in learning about actual assembly programming, I recommend Microchip's line of PIC processors. Different companies sell very reasonably priced prototype boards to experiment with, and since the PIC line has been around a long time, there are any number of books on the subject. I urge you to go for it. It's fun.[13]

∞

You're probably thinking, oh geez, he's finishing yet another chapter, and he *still* hasn't explained how a binary on/off operator can execute program instructions to change your home's temperature via your phone. I know, and I do

---

[12] Hexadecimal is a way of showing binary values in more human-readable form. We group the binary field into four-bit sections (called nibbles), and each section is then converted into a decimal number, where the values above 9 are given alphabetic letters. So, 00010101 (0001_0101)becomes hexadecimal 15, and 11000001 (1100_0001) becomes C1 (1100 is 12, which is "C").

[13] By the way, this small instruction set I've made up for demonstration purposes could actually be compiled to implement 99% of any high level program. Most of it would be quite tedious, however—like writing this entire book with just 100 words (which you may wish I'd done). But, more important than making the job easier, larger instruction sets require less steps, resulting in faster operation. And isn't that what it's all about?

apologize. But what I have shown you so far is maybe more valuable—that computers are hugely complex, and you should have more appreciation for the engineers who design and program them.

Also, we're not done. Hang in there.

16

Proper Protocol

The Internet

Using your imagination and what you've learned so far, you might have an idea how the computer in your phone could watch the keypad or screen for indications of directions from nimble fingers. It sees you select the application (app) for remote access and control of your smart home environment control system.[14] The operating system calls up the app, launches it, and steps aside (still watching over the app's shoulder to make sure it behaves). The app takes your nimble finger inputs, asks the operating system if you're currently connected to the internet via WiFi or L4 service, and if not, it convinces the operating system to make a telephone call to your service provider, where the internet connection is made. The app then initiates an internet connection to your home WiFi, which routes the connection to your smart thermostat.

But, what exactly is the internet?

As you probably already know, the internet was invented

---

[14] Within the culture that exists as I write this, I feel it necessary to indicate when I'm talking about electronic devices with enough enhanced computer processing to earn them the title of "smart." Future generations who read this, and I can only hope there will be, may find this quaint or even confusing, since for them a "smart" device may be one that can help them with their homework. Sorry.

by Al Gore in 1992. Sorry, I couldn't resist.[15] The internet was actually developed by DARPA, the government defense agency created by President Eisenhower to explore the boundaries of technology, looking for better ways to beat his rival, Nikita Khrushchev. The original DARPA network was called ARPANET, and went online in 1969, connecting three universities and a private research center. Over the next decade it grew, adding additional academic-oriented locations, and by 1981 it included 213 connected computers. In 1983, the ARPANET was split—a smaller portion snipped off to be used exclusively for the military, and the rest, expanded by the National Science Foundation as a sort of foster parent, became what we know as the modern internet. The NSF bowed out in 1995, leaving the internet to find its own private/commercial way, either fulfilling the final promise of ultimate human interconnection, or one huge bucket-load of trash, depending on your perspective (and maybe age).

The defining components of the original ARPANET that established the basis for today's internet were 1) a working implementation of packet switching, and 2) the use of layered protocols to manage the switched traffic. I imagine that these are all words you've heard before, but let's open the flaps and take a peek inside.

Packet switching is a concept that takes advantage of the discrete nature of digital information, a method of communication impossible with analog signals. The basic idea is pretty simple. Prior to packet switching, the data—both analog and digital—was carried from one location to another across a fixed "circuit." A circuit was a dedicated connection between two locations, for example a telephone connection made via some number of switch locations across the country. The dedicated connection may only exist for just as long as the data is transferred, i.e., it could be set up and then torn

---

[15] What Al Gore actually said in a CNN interview was that, while in Congress, he initiated legislation that fostered the budding internet. Gore never claimed to have invented it, and the accusation that he said he did was an invention of perhaps equal audacity. Al Gore did coin the term "information superhighway."

down; the component links could then used for other circuits.[16]

With packet switching, a message—maybe an email, maybe a gigabyte file—is broken up into a bunch of small pieces, which are then sent out into a mesh of general connecting links (the internet's physical infrastructure). Each piece (called a packet) has its own copy of the destination address—the address of the computer we're trying to get the message to—and finds its own way there, shuttled along from node to node, each node moving it along closer to the destination address.

This probably sounds a little higgledy-piggledy, sort of like tossing all the sheets of your three-ring binder into the wind and hoping your friend in the next dorm will collect them all and put them back in order. But if computers are good at anything, it's dealing with seemingly endless repetitions of complicated tasks. The advantage of this sort of communication is that it maximizes the use of the physical network. (It's sort of like transporting a dozen bird cages using the trunk of your car. Carrying them intact, you'd have to take half a dozen trips. But if you grab a screwdriver, disassemble them, tag each piece, and pack them in snugly, you could do it all in one trip. You've eliminated the empty spaces inside the cages.)

What I've just described is the fundamental core operation of the internet. This was the basis of the original ARPANET. But the modern use of the internet includes other standardized layers as well. It's the combination of these layers that allows your computer (or phone) to connect and transfer information, such as pictures of your niece, or your monthly electric bill, or, yes, instructions about how to set the temperature of your house using your phone. Each layer has rules that are followed, called protocols. Driving around

---

[16] This is the simplest form. For the long-haul links—those between major regional switching nodes—the physical wire that carries the data may be time-sliced, allowing multiple circuits to simultaneously share that wire. But the slices of periodic time are themselves dedicated to allocated circuits, so there exists a complete dedicated channel between the two end connections.

Manhattan, every driver had better turn in the correct direction on one-way streets, and stop and wait at red lights—in other words, follow the necessary protocols—for the city to avoid complete gridlock (and crashes).[17]

The core packet switching layer described above is called the Internet Protocol—"IP" for short. It's the lowest layer defined in the suite of standardized internet layers. There are layers below it—specifically a variety of different "link" and "phy" (physical) layers—that are responsible for getting the IP packets from one switch node to another, but the IP packets don't care what they are, as long as they deliver, just as you don't care what the roads under your car tires are made of—asphalt or concrete (okay, you may care, but let's not get picky). The packet-switching address of the destination computer is called, appropriately enough, the IP address. This is exactly the same IP address your internet service provider help-desk person bandied about when getting you connected.

With a physical mesh, or network, of connected spans (the link and phy layers), network nodes that understand how to forward IP-structured packets, and connected computers that can chop up and attach IP addresses on outgoing messages (packets), you have a working internet.

Sort of. What you have so far is akin to saying that since we have the UPS company, and a network of highways and streets, your father can take apart the playground equipment you used as a kid, and send it to you for your own kids to use. UPS will drop a hundred boxes at your doorstep, and the job of our metaphorical internet is done.

What we need to re-assemble the jungle gym and swing set are protocol layers above IP. There are various higher layers than can ride on the internet (IP), but TCP (Transmission Control Protocol) is by far the most common. This layer is responsible for making sure all the individual

---

[17] And the reason that Manhattan still suffers from chronic gridlock (and crashes) is that there's not enough bandwidth—i.e., too much "traffic," for the available channels (streets).

packets make it to their destination error-free, and then are re-assembled again.[18] TCP is an end-to-end protocol. Once a packet is sent on its way from each node, the job of the IP layer is done. TCP keeps an eye on both ends—if one of the packets goes missing (e.g., is dropped due to congestion), or arrives errored, it's TCP's job to re-transmit that packet. TCP's history goes back to the beginnings of the internet and it comprises so much of the use of the internet, that it is usually coupled with the IP layer, with the two referred to jointly as TCP/IP.

With both IP and TCP, we now have the means to move complete messages (e.g., a picture) from one computer to another. We're finally ready for user-familiar application programs. But we have to pause a moment, and explain the World Wide Web. This is the "www" at the beginning of the names of sites that you go to on the internet—"www.readler.com," for example. Many people equate the internet with the "web," but the World Wide Web is just one use of the internet, albeit the predominant one, by many factors. The World Wide Web is not an application, but a place. This "place" consists of millions of individual places, and these are, of course, web pages. Your browser is the software application that manages your visits. Note that a browser is a software application, but it's not *the* protocol of the application layer of the internet protocol layer structure. It can be confusing. Your browser uses two different internet application *layer* protocols—DNS and HTTP. DNS stands for Domain Name System, and it's what equates the web names you're familiar with—e.g. "www.readler.com"—with IP addresses, which, as we know, is how the browser actually connects with the page on the internet (more accurately, to the server computer hosting your web page). HTTP stands for Hypertext Transfer Protocol, and this defines a way for your browser to request the information contained on the web page from the computer hosting it. The information is usually in

---

[18] TCP is one of the protocols of the transport layer. So, we have, from bottom to top, the Phy layer, the Link layer, the Internet layer (IP), and now the transport layer (TCP).

the form of HTML files, which are text-based files with extra embedded information that tells your browser how to display it.

So, for example, if you type "http://www.readler.com/" in the web address window of your browser, you're effectively telling the browser to use its HTTP protocol to go to the IP address associated with www.readler.com and see if there's anything to display. In this case, the browser finds an HTML file that displays a very handsome and informative suite of goodies.

Your email program (properly called an email client) is another example of a software application *program* that uses internet application *layers*. Here, the two predominant internet application layers are POP (Post Office Protocol), which defines how your computer communicates with your email service server, and SMTP (Simple Mail Transfer Protocol) which defines how the emails themselves are structured and communicated.

Another common internet application layer is FTP (File Transfer Protocol), an ancient protocol whose beginnings are coincident with ARPANET itself. As its name implies, it is a streamlined means of transferring files between a host and client computer. Note that you type "ftp://" instead of "http://", since these are two different application layer protocols.[19]

All this is fine, but I haven't explained how Ethernet and WiFi fit in. Ethernet existed in parallel with the internet through the 1980s and much of the 1990s, and was (is) a means to network together co-located computers, thus the term LAN (Local Area Network). LANs (i.e., Ethernet) are what typically tie together computers within a company. In the 1990s, most users (e.g., me) connected to the internet via a telephone modem. If we were also connected to a LAN, the Ethernet port of our computer was usually dedicated to that. Once we found better (faster) means of connecting to the internet, specifically via our cable company or DSL lines from

---

[19] Actually, since we use FTP to transfer files, we often use a file browser, such as Windows Explorer, instead of a web browser.

our telephone company, we needed a method to interface with the cable modem or DSL termination. Since Ethernet was an established link method, that interface port on our computer was pressed into service. Ethernet carries us to our ISP (Internet Service Provider), which then connects us into the internet. So, until we get tapped into the internet, Ethernet serves as both the Phy and Link layers of the hierarchy. Our computers load the IP packets (which include the TCP and application information) on top of Ethernet "frames."

WiFi is used as an alternative to the wired Ethernet, but serves the same purpose, getting us to our ISP. WiFi, though, is a relative new kid on the block—born at the beginning of the millennium—and has never been used for much else other than internet access.

Finally, a word about routers. These were originally used to interconnect different networks, and, as such, were substantial computing platforms that you and I (okay, you) would never see, being located in large air-conditioned rooms that you needed a special badge to enter. Their specialty is handling disparate network formats, receiving and forwarding packets with different network protocols (a good example is connecting an Ethernet LAN to an internet backbone). Routers were also switches. They analyzed the addressing of the different formats and not only translated the format, but sent the packet off in the correct direction. Initially big, costly machines, through miniaturization and adaptation to our growing need for small scale internet access, routers shrank both in size and in functional variety, until they now sit on our shelves blinking their LEDs to tell us that our interconnection is up. Our cable modems and WiFi boxes are indeed routers, but with limited switching functions, and only able to convert the specific packet formats between our computer and our ISP.

∞

For eighty years the telegraph ruled as the technological wonder connecting the human race. For the next eighty years, it was the telephone. Now it's the internet. If science and technology continue the pattern, we should be looking for the next communication means (electronic telepathic emulators embedded in our heads?) around 2080.

I'll update this book with a new chapter when the time comes, so watch for it.

17

A Mishmash of Nuggets

This is a bunch of stuff that didn't fit anywhere else. Not impressive structure management, but maybe better than not including it at all.

Microwave oven—boiling an egg over the radio

A microwave oven heats food by blasting it with radiation, and, as you know, radiation causes materials to become radioactive, so all the microwave-warmed food you ever ate has inflicted extensive DNA damage throughout your body, causing an untold number of cells to transition to a cancerous state. You are doomed.

I remember exactly this sort of hand-wringing worry when my grandfather bought an Amana Radarange for my mom in 1972 (an RR-4 model, with about twenty pounds of chrome). I hadn't started my engineering studies yet, and so I was open to the idea that my grandfather had unwittingly brought tragedy down upon the heads of his own brood. On the other hand, in my imagination it was brand new technology[1]—possibly spun off from NASA's Apollo program (it wasn't)—and way too cool to avoid. I spent hours experimenting with different substances until my mom put a stop to it when, with a heart-stopping boom, I exploded an egg. Despite my sustained close encounters with the invisible energy, I still live to tell of it.

In fact, microwave oven radio energy is in the 2.4 GHz range, the same frequencies used by cordless phones, garage door openers, remote controlled toys—virtually any radio-based remote operation. It's just a lot stronger. And more contained. The important point is that it's not ionizing energy, meaning that it can't knock electrons out of the atoms it encounters. If it did (which it doesn't), this would cause atoms to become ions, and these in turn can cause damage to cellular molecules, like your DNA, or even damage your DNA directly. This is the danger you hear associated with free radicals and real atomic radiation—the kind that results from nuclear reactions. If you defeated the door interlocks and managed to stick your hand into a working microwave oven, it would feel warm, and then quickly painfully hot. Assuming you didn't leave it there long enough to cook yourself, you would suffer no permanent damage. The psychological damage that would follow when you realized how stupid that was is a different story.

Water is a polarized molecule, meaning that, although it is overall "charge-neutral," at the molecular level it's unbalanced, i.e., it has a positive side, and a negative side. Many molecules have some amount of polarization, but water's is pronounced.[2] In a microwave oven, the energy is electromagnetic, as is AM/FM radio, light, x-ray, or infrared.[3]

---

[1] Monster commercial microwave ovens—also trade-named Radarange—had been around since the 1940s. At that time, microwave frequency generation was new, created as part of the radar development during the war. Thus the now-archaic name.

[2] The negative side of one water molecule attracts the positive side of another. Thus, water molecules tend to stick together. This is what surface tension is all about.

[3] Whereas microwaves are not ionizing, X-rays are. This is why we're told to limit the number of CT scans we get (which use X-rays). Whether a certain electromagnetic energy is ionizing or not depends on its inherent energy, which is proportional to its frequency.

Note that "magnetic" is part of the word. The positive and negative sides of the water molecule align themselves to the magnetic aspect of the microwave energy. However, this energy is oscillating billions of times each second, which jiggles the water molecule (really, really fast). The jiggling molecules bang into other molecules (like meat protein) and set them moving. This is heat. Note that no electrons were harmed in this exercise.

The microwaves are about five inches in length, and that's why you can look into the chamber to see if your oatmeal is boiling over. The mesh holes are too small to let them through (the microwaves, not the oatmeal). Most of the radio waves generated by the magnetron tube don't directly hit your food. Instead, the energy just bounces around inside until it finally stumbles into it. The inside of the microwave oven is carefully designed to maximize the bouncing around— to avoid hot or cold spots. If the energy doesn't find a food target, it could end up synchronizing with later generated waves. This is called a standing wave, and overheats the magnetron tube. After all, the energy has to go somewhere (the first law of thermodynamics: energy can be neither created, nor destroyed).

By the way, don't try cooking a whole egg in your microwave. The mess is astounding, and you probably don't have a mom to clean up after you.

Air conditioners and heat pumps

We mentioned these back in chapter 4, but to complete the picture, a window air conditioner is essentially a small refrigerator manhandled into a window, facing inward with the door removed. You then put a portable fan inside to draw the cold air out. The difference between a bona fide air conditioner and a small refrigerator is that the air conditioner

---

Note that as we go down in frequency (and thus energy level) from X-rays, we go through ultra-violet (ionizing at the higher end), visible light, and infrared, before we come to microwaves. Based on this, you should be more worried about your toaster oven (infrared) than your microwave.

is made to fit the window, and it has a larger compressor motor (and coil sizes to match), consequently using more power.

If you have central air conditioning, then the compressor motor, heat exchanger, and giant exhaust fan are contained in one big, louvered box, located outside somewhere where the roar bothers your neighbors instead of you, and the evaporator coils are (usually) placed inside your central heating system, where that fan serves double-duty. The refrigerant fluid/gas is piped back and forth via thick, insulated pipes.

If you have a central heat pump, then it's essentially the same, except that there's a valve in the compressor that can reverse the flow of the refrigerant. When the flow is reversed, what was the evaporator coils (where the gas was allowed to expand) inside the central heating ducting becomes the condenser, and what was the condenser coils in the outside heat exchanger now becomes the evaporator. After all, a pipe that is big and gets smaller (to condense the gas) in one direction, is a pipe that is small and gets bigger (to expand, or evaporate the gas) in the other direction.

They're all refrigerators.

Inductive stoves

These appliances are essentially powerful transformers with the secondary coil removed, and the iron core replaced with your cooking pot. The difference here, though, is that your pot (the transformer's core) is soaked with so much magnetic energy that it gets saturated, and after that, the energy is converted into heat. Directly in the bottom of your pot. These stoves are the most efficient way available to cook food, and they have the same advantage as gas cooking in that the heat is present almost instantly. Also, since the heat is delivered directly to the pot, the stove's surface is only as hot as the pot—here, the heat's going backwards.

The downside is that you will probably have to throw away much, or maybe all, of your present cookware, since induction heating only works with cast iron or some stainless steel. Also, the pots and pans need to have nearly perfectly flat bottoms. Essentially, you should probably plan on replacing your entire set of pots and pans with ones made specifically for the job.

Power savor capacitors

Both inductive loads (e.g., motors) and capacitive loads (e.g. . . . um, nothing of any consequence that you would probably have in your home) affect something called the power factor in the electricity being supplied by the power utility company. The power factor is an indication of how much out of phase the voltage and current are with each other. This is quite an esoteric subject (I didn't understand it until my senior year in college), but suffice it to say that controlling it is a major headache for the utility company. The inductive loads of large motors of industrial facilities create so much power factor (reducing it) that the power companies require them to install compensating capacitor banks. The power company then installs a power factor meter to make sure the result stays within bounds. The facility gets fined if not.

Now, your measly little house has a few motors that run intermittently, and you do create small power factor offsets. The power utility would be pleased if all homeowners could compensate for that, but it's just not possible. To properly compensate, you'd have to vary the amount of capacitors in conjunction with all our little motors coming on and off. That's obviously not feasible. In fact, if every homeowner installed one of the fixed capacitors that are currently being marketed, it's likely that the combined power factor might go the other way, and the power utility won't stand for that. They'd require us all to remove them.

These power-saving capacitors are advertised as devices that save you money on your electric bill. They claim that without them, the power factor offset causes your meter to run faster. That's just not true. You are billed for what's called the "real power," while the power factor affects what's called the "apparent power." The apparent power can only be equal to (with no inductive or capacitive loads at all) or bigger than the real power. It's the apparent power that the utility needs to generate, so you're actually getting a good deal off the bat.

In fact, these power-saving capacitors actually draw current of their own, so if anything, they're costing you more.

If you're not convinced, and are determined to buy one from the friend of a friend who's friended you on Facebook, then you should talk to me. I have invented this device that records your telephone conversations, and after you disconnect, it makes a dummy call and plays the whole conversation backwards. This reverses all the bits, and eliminates your call from the phone company's records.

Batteries

Batteries not included—the bane of parents at Christmas time. If batteries do happen to be included, they're rarely alkaline, and I would often throw these away and grab a couple of alkaline from the stock I keep. This was based on my long-held belief that alkalines were always longer-lasting than the regular old zinc-carbon batteries of my youth. In applications that require a constant low draw over a long period, alkaline batteries beat the older kind hands down, lasting four times longer or more. An example of this is the remote temperature sensor that's hanging outside my window. Alarm sensors are another.

But alkaline batteries only win races by being the slow plodding turtle. They don't do well with short, large-draw applications. In this case, they do no better than the "heavy duty," or "super heavy duty," kind.[4] Examples of applications

---

[4] There are two kinds of non-alkaline (non-rechargeable) batteries. The older kind, going back to the fifties, are zinc-carbon. These are usually the super-cheap kind that come as generic brands. The newer kind are zinc-chloride, and are usually labeled with variations of "heavy," "super heavy," "extra heavy," etc. Despite the variety of extravagant monikers,

of this kind are digital cameras and, yes, remote-powered toys. In these cases, since the heavy-duty batteries typically cost less than half of the alkaline, they'd be the better choice. So, there's some irony in the fact that both Duracell and Energizer chose rabbits for their commercials.

Actually, short, large-draw applications are better suited to rechargeable batteries. Although they cost two or three times as much as alkaline, they have comparable capacity— greater, even, for the large-draw modes. Just three recharge cycles, and you're ahead of the game. They fall down in longevity, however. They self-discharge at a high rate, meaning that just sitting on a table, they'll go dead in a few months (down to 30% of full charge after three months). They're good for up to a month or so, though. Of the two types, the NiMH have two to three times the capacity of the older NiCad, but also have a faster self-discharge (30% per month, versus 20% per month for the NiCad). Neither should ever be allowed to go completely dead, nor be over-charged. Both of these conditions damage them, and they shorten their capacity and life. For example, you might not want to use them in a flashlight, which can be accidentally left on, and which will then drain them dead. Otherwise, they're good for hundreds of charge cycles.

One minor problem with rechargeable batteries is that their fully charged voltage is 1.2 volts, versus the traditional alkaline and "heavy duty," which are 1.5 volts. Newer devices are designed with this in mind, and most older ones work okay at the lower voltage, but if you have problems using a rechargeable, this may be the reason.

Also, whereas the voltage of alkaline and "heavy duty" batteries fall at a fairly constant rate as they're drained,[5] both types of rechargeables have the unwieldy habit of holding steady at their 1.2 volts until they're just about dead, and

---

they're all the same, and they have two times or more the capacity of the cheapo zinc-carbon.

[5] Until about 1.0 Volts, after which, the voltage drops precipitously, or it would, but at this point, your device has stopped working anyway.

then dropping dead. I use rechargeables in an old portable CD player, and, when inserted, freshly charged, the player indicates that they're almost done—this is because this is the case when an alkaline battery falls to 1.2V. I use the player for weeks, and the battery indicator continues to read *almost dead, almost dead, almost dead.* And then, without warning, the player just stops cold.

Here's some comparisons you might find useful:

• AA batteries have about twice the capacity of AAA; C size have about seven times as much; and D size have about fifteen times as much. Remember the 6-volt lantern batteries? They have nearly eighty times as much;[6]

• compared to alkaline, zinc-carbon batteries (the original regular type) have about 40% the capacity, zinc-chloride (e.g., "super heavy") about the same, NiCad about half, and NiMH about the same (newer ones, even more);

• rechargeable batteries can deliver more instantaneous current than alkaline;

• lithium batteries cost about four times as much as alkaline, and have two to three times the life. They have a much longer shelf-life, though, and work better at cold temperatures, so these are good choices for remote, outside applications (e.g., my temperature sensor).

Lightning Rods

If your house stands alone—no tall trees next to it—you should probably have one.[7] Lightning typically thrusts over 50,000 amps at you, and without a lightning rod, it struggles as it finds a path down through your structure. The high current combined with the high resistance of your wood frame means a whole lot of heat. So much that where the lightning

---

[6] The 80x factor for the 6-Volt lantern battery takes into account that the voltage is four times that of the AA, i.e., the capacity here is the available energy.

[7] Unless you live in southern California. Thunder storms are rare enough that people come outside to enjoy the show. These are the same people that don't even get up from their desks when an earthquake strikes unless things start falling onto the floor.

strikes, it seems as though that part of your house explodes. Because it does.

Lightning protection systems are simple as pie: a pointy rod sticking up higher than anything else (multiple if you have a long ridge), with large diameter cable feeding down to a second grounding rod. The two rods can be pretty much anything that's made of metal—old rebar, or the lever that swings up to prop up your car hood. You can't use any old lamp wire for the connecting cable, though.[8] This should be a half inch or more in diameter. It looks like the cables they use to moor supertankers. You can buy this stuff online from specialty companies, and a hundred feet will cost you a couple of hundred dollars or so.

An interesting note. As you probably know, Benjamin Franklin invented the lightning rod, and soon after, the unborn United States and Father England had a falling out. Fists swung, musket balls flew, and everybody took to their own sides, including scientists. The English were sure that rods topped with balls were the ideal lighting attracter, while the Americans followed Franklin's lead and swore that pointed rods were the bees knees. To this day there is no agreement,[9] and all over England you see ball-topped rods in the air, and in the United States, rods with pointed ends.

Laser pointer

All I have to say here is that if we ever invent a time machine, sneak in after hours, take one of these back to 1967, and then flick it around casually and remark, "Oh, yes. We have these in the future. Only secret agents get to carry one.

---

[8] Let's do the arithmetic. Say the lamp cord exhibits 0.1 Ohms from roof to ground. Power = I^2 x R. 50,000 Amps squared times 0.1 Ohms equals 250 million Watts. The lamp wire will instantly melt, and as it breaks apart, the lightning will jump the gaps, ionizing the air. Within a couple of microseconds, the wire will be replaced by a path of ionized air, which the lightning likes. If you're lucky, the lightning will then just scorch the outside of your house where the wire used to be. If you're not lucky, your house will still burn down.

[9] After all, it's not easy generating 100,000 Amps in the lab to experiment.

Such as I." Don't mention, though, that our bridges are crumbling and we're taking pictures of our ice caps to show to future generations.

Index

www.ingramcontent.com/pod-product-compliance
Lightning Source LLC
Chambersburg PA
CBHW050507210326
41521CB00011B/2355